# Significant Concepts of Biodynamics, Biodiversity and Soil Fertility in Agriculture

# Significant Concepts of Biodynamics, Biodiversity and Soil Fertility in Agriculture

Edited by **Jordan Berg**

SYRAWOOD
PUBLISHING HOUSE

New York

Published by Syrawood Publishing House,
750 Third Avenue, 9th Floor,
New York, NY 10017, USA
www.syrawoodpublishinghouse.com

**Significant Concepts of Biodynamics, Biodiversity and Soil Fertility in Agriculture**
Edited by Jordan Berg

International Standard Book Number: 978-1-68286-008-3 (Hardback)

Printed in the United States of America.

# Contents

# Preface

This book has been an outcome of determined endeavour from a group of educationists in the field. The primary objective was to involve a broad spectrum of professionals from diverse cultural background involved in the field for developing new researches. The book not only targets students but also scholars pursuing higher research for further enhancement of the theoretical and practical applications of the subject.

Different approaches, evaluations, methodologies and advanced studies on agriculturally significant topics, such as soil classification, fertility and development; seed science, agroecology; biodynamic and organic agriculture; etc. have been included in this book. The aim of this book is to present researches that have transformed agricultural practices and aided its advancement. Scientists and students actively engaged in the field of agricultural sciences, plant science and associated disciplines will find this book full of crucial and unexplored concepts.

It was an honour to edit such a profound book and also a challenging task to compile and examine all the relevant data for accuracy and originality. I wish to acknowledge the efforts of the contributors for submitting such brilliant and diverse chapters in the field and for endlessly working for the completion of the book. Last, but not the least; I thank my family for being a constant source of support in all my research endeavours.

**Editor**

# Strategy for Insect Pest Control in Cocoa

**Richard Adu-Acheampong[1*], Joseph Easmon Sarfo[1], Ernest Felix Appiah[1], Abraham Nkansah[1], Godfred Awudzi[1], Emmanuel Obeng[1], Phebe Tagbor[1] and Richard Sem[1]**

*[1]Entomology Division, Cocoa Research Institute of Ghana, P.O.Box 8, Tafo-Akim, Ghana.*

***Authors' contributions***

*This work was carried out in collaboration between all authors. Authors RAA, JES, EFA and GA defined the research theme and wrote the first draft of the manuscript. Authors AN, PT and RS co-worked on associated data collection and their interpretation. Author EO provided additional information from field experiments. PT reviewed all drafts of the manuscript. All authors read and approved the final manuscript.*

Editor(s):
(1) Marco Aurelio Cristancho, National Center for Coffee Research, CENICAFÉ, Colombia.
Reviewers:
(1) Anonymous, Greece.
(2) Anonymous, South Africa.
(3) Anonymous, India.
(4) Anonymous, Ghana.
(5) Anonymous, Italy.
(6) Ian Mashezha, Zimbabwe Open University, Zimbabwe.

## ABSTRACT

Farming systems in cocoa over the last three decades have involved the use of new hybrid plant varieties, which produce pods throughout the year, intensified fertilizer use, and misguided pesticide applications by some farmers. Resource availability in terms of abundance of feeding and breeding sites and ecological disruption as a consequence of climate change and bad agronomic practices have increased the importance of insect pests on cocoa. Historically the major management tool for hemipteran pests has been calendar spraying with conventional insecticides. Considerable progress was made at the turn of the last century by replacing organochlorine insecticides for cocoa mirid control. But inappropriate timing and inefficient application is probably reducing the viability of otherwise acceptable products in some areas. Integrated Pest Management (IPM) strategy for mirids and other insect control should involve great investment in pest surveillance, and be based primarily on the use of cultural practices of removal of excess chupons, shade management and

---

*\*Corresponding author: E-mail: r.aduacheampong@yahoo.co.uk;*

host variety resistance. These practices must primarily aim at minimising mirid-associated dieback disease and stink bug deformation of pods, and should be supplemented in some cases by the use of insecticides (up to two applications in February to May at 28-days intervals) depending on the pest populations, damage levels as well as intensity of activity of pollinating insects, with additional two applications during September to December when pest problems may arise. Improved methods of monitoring and prediction should assist in early identification of specific problems in different farms. The rotational use of different active ingredients should also take into account factors such as application methods, compatibility and correct timing. Careful planning is necessary to formulate a flexible control system.

Keywords: Mirid; pest control; cocoa; farming systems; control strategy; correct timing; dieback disease; insect pest; flexible control system.

## 1. INTRODUCTION

Mirids (*Distantiella theobroma* [Dist.] and *Sahlbergella singularis* Hagl.) and stink bugs (*Bathycoelia thalassina* (Herrich-Schaeffer) are known for their importance in causing losses in cocoa (*Theobroma cacao* L.) in West Africa [1-3], with the most prevalent in Ghana being *S. singularis*. The authors have outlined the problems associated with mirid and stink bug infestation. Both nymphal and adult stages feed and inject saliva to liquefy plant tissues. Their feeding punctures provide entry points for fungal pathogens [4-7]. This secondary invasion of mirid feeding punctures causes dieback disease whilst that by stink bugs results in premature ripening and deformation of pods and beans. The main method of mirid control is by the application of conventional insecticides, but stink bugs have often been controlled indirectly by treatments normally used for the control of mirids.

In Ghana, new pests have emerged and their control is difficult compared with three decades ago [8]. There are several contributory reasons for this: farmers still have to be familiarized with optimal techniques of using the new generation insecticides that have largely displaced the organochlorine and carbamate compounds. The global quality standards of insect pest control in cocoa needed to guarantee high quality produce for global market outlets have increased, setting targets for maximum residue limits of pesticides, which in some instances are clearly not attainable.

A fast and correct diagnosis of pest incidence is the first important step in attempting any control procedure. Earlier studies [9] have concluded that pest control systems in cocoa must be revised as a result of changes in farming systems, pest cycles and shifts in pest population peaks.

Before discussing the strategy for mirid control in cocoa in Ghana it would be useful to examine the history of this insect's control.

Before the 1970s damage caused by mirids was well appreciated as well as the importance of dieback disease that follows mirid feeding on green unhardened tissue (pods and stem). Efforts were concentrated on understanding population fluctuations of mirids on different varieties of cocoa, their distribution within trees as well as the effect of water stress on the incidence of pests. The tendency for average population levels to remain low until a sharp rise in August or September was the basis for the establishment of the calendar spray regime beginning in August until December (but omitting November) [10]. This was to replace an older recommendation of June, July followed by a three-month gap [11]. Currently, two seasonal peaks varying from September-December and February-May in different locations is an average phenomenon. Plant breeders at the Cocoa Research Institute of Ghana (CRIG) have developed varieties of cocoa resistant to swollen shoot and black pod which are also high yielding but unfortunately this resistance does not seem to be durable enough when the varieties are challenged under farmer conditions. Very little effort was spent on breeding varieties with resistance to mirids or other insect pests. Field observations usually gave strong indications that mirids caused less damage to some Trinitario selections (e.g. SC1) than the common West African Amelonado [12]. This was confirmed later [13], with SC6 identified as one of the tolerant selections to fungal-induced dieback disease in cocoa. In compiling its list of recommended varieties to the seed gardens, CRIG takes a firm line in rejecting varieties susceptible to swollen shoot disease and encourage the development of resistance to black pod disease. Unfortunately, the Institute does not follow an equally firm policy for insect pests.

Insecticides (nicotine sulphate and DDT) were first introduced in the 1940s against mirids because accumulated knowledge highlighted these insects as the most important pests. In the following years, several insecticides including endrin, dieldrin, heptachlor and lindane were introduced against mirids, which were also aimed at the control of mealybugs (*Planococcoides njalensis* (Laing), *Planococcus citri* (Risso) and *Ferrisia virgata* (CkII)) which are vectors of cocoa swollen shoot virus disease. Unfortunately, insecticides were applied prophylatically and at times when the pests were already present at damaging levels.

The percentage of cocoa trees that was treated with insecticides in the 1950s is difficult to estimate accurately, but it rapidly increased in the 1960s [14]. Cocoa output reached an all-time high of 560,000 tonnes in 1964/1965. Nearly all the insecticides used were against mirids due to their relative high economic importance while the adoption of insecticide use for other insect pests was slower. Later on, a growing concern for the environment and for the conservation of wildlife was raised. At present, environmental and consumer safety dictate the sustainable usage of chemicals. In this connection cocoa farms are unique because of the large areas they occupy (nearly 2 million hectares in Ghana) which is a major factor to be taken into account when considering strategies for pest control.

During the last two decades, and especially since 2000, there have been very significant developments in nearly all aspects of cocoa insect pest control. Farmers have become very conscious of the importance of insect pests, and data accumulated in mirid surveys on cocoa would enable objective assessments to be made [9]. Although data on yield and financial losses due to pests are not readily available (due to varying farm management systems), surveys demonstrated clearly that mirids and stink bugs are the most important pests [15].

The field of mealybug-vector control has seen fewer successes. The use of systemic insecticides such as Monocrotophos SC was a much more convenient way of controlling swollen shoot virus vector in cocoa. The method suffered a setback when it was detected that the chemicals (mainly organophosphate insecticides) caused tainting of beans [16].

For control of pod feeding insect pests, the present system of insecticide usage ensures that pests are kept at very low levels. A possible danger now is that famers faced with increased insecticide costs have tended to procure cheaper unapproved insecticides and omit the registered insecticide treatment. This calls for an urgent need to make the approved insecticide products sufficiently available (in pesticide retail shops) so as to cause a reduction in prices through market competition.

Improvements in the monitoring and forecasting of the major pests on the new varieties of cocoa recently became evident [17], but the shift in dominance between the two important mirid species is not adequately understood, and should be studied further. Cultural practices which reduce serious insect pest attacks should obviously play an important part in a control system. Such practices may include pruning, chupon removal, agro-forestry, shade management, removal of alternative hosts, weeding and fertilizer application [18]. Apart from these, the main emphasis in insect pest control will be on the use of resistant varieties reserving the use of chemicals for when economic threshold levels are exceeded.

## 2. BIOLOGICAL CONTROL

Past observations in Sierra Leone indicated low levels of mirid infestation. This prompted a search for natural enemies in Sierra Leone and the Congo but no significant effort to introduce natural enemies have thus far being done.

## 3. HOST PLANT RESISTANCE

Nearly all the damage to cocoa trees of the 1950-1960s was associated with the growing of very susceptible varieties including Amelonado [19]. But no variety has expressed a satisfactory level of resistance to *S. singularis* and *D. theobroma*. In one of the few studies to address the issue, [20] found that many local and international cocoa selections were highly preferred by mirids as suitable hosts for feeding and oviposition. This apparent lack of genetic variation in host preference requires further investigation, especially since genetic variation for resistance has been observed in several other systems [21,22]. However, the development of a strategy for pest control could include varieties resistant to dieback disease resulting from hemipteran related fungal invasions as was shown in later studies [13].

Identification and categorization of sources of resistance for developing pest resistant cocoa varieties should be pursued by biotechnology and traditional plant breeding. Observations at CRIG have shown that by mixing several varieties (each of them contributing a different form of genetic resistance), pest outbreaks can be reduced [23]. The push-pull effect through use of semio-chemicals to repel insect pests from the crop ('push') and to attract them into trap crops ('pull') must be exploited in future IPM strategy. Farmers seem to have understood the need for cocoa varieties as well as shade trees and food intercrops. It will be interesting to see whether this work will lead to at least a partial solution, of the cocoa mirid problem.

## 4. MONITORING AND PREDICTION

Hemipteran pest populations in cocoa are spatially and temporally patchy with many individuals exploiting cryptic habitats [24]. An ability to predict or forecast severe outbreaks of pests is obviously an advantage when planning control programmes as that can lead to the selective, timely and precise use of chemicals.

Information from farms can be collected and fed into a central computer database, which can be managed by CRIG and the Cocoa Health and Extension Division of Ghana Cocoa Board. These organizations can receive information from many sources apart from their staff, notably the representatives of the many licensed cocoa buying companies and certification agencies serving farmers, as well as directly from farmers themselves. This information can be processed and released at fortnightly or monthly intervals for use on radio, television and the press. Farmers can receive useful guidance on the pests they should be looking out for. An improvement in the use of information dissemination services would lead to timely awareness of insect pest problems in cocoa farms as well as drastically reduce response times.

There is, however, very little information which can be used to predict the spread of pests. Whilst it is unlikely that spraying for insect pest control could be based entirely on prediction systems, a planned programme of timely sprays could be modified if the status of the biology of the specific insect pest was known. For example, that certain weather conditions might be favourable for the spread of a particular insect pest. Moreover, it is suggested that a monitoring programme where pest incidence as well as meteorological parameters are monitored should be established.

## 5. CHEMICAL CONTROL

Insecticide use in cocoa in the future should aim to supplement the control given by cultural practices, varietal selection and other methods, within an Integrated Pest Management programme. Experience gained from pest assessments at CRIG may lead to timely chemical treatments of cocoa which is sustainable in the case of cocoa pest control. It is common for farmers who have invested a lot of time and money in maintaining their cocoa farms to spray prophylatically as a precaution against possible pest outbreaks [9]. It is also true that it is difficult to predict the development of some pests, such as stink bugs and stem borer adults in cocoa. Farmers may also be unable to spray a large area at short notice. Furthermore, a high yielding cocoa farm is more likely to be susceptible to insect pests than a low yielding cocoa farm (as there would be plenty of feeding and egg laying sites [9], and with a higher potential profit the farmer can afford treatment costs. There should, thus, be separate timely recommendations for potentially high yielding cocoa farms as routine sprays on cocoa farms with lower than average potential yields are generally uneconomical.

Where insecticide use is planned, the approach should be flexible, taking into account the likely occurrence of pests but will also depend on factors such as geographical situation, varietal choices, and weather at critical times, as well as facilities available to the famer to apply the chemical.

In recent years, insecticides registered for mirids have been used in more than 40% of cocoa farms in Ghana. Discussions with farmers suggest that the majority of the unsprayed cocoa farms would also have benefited from the government's 'mass spraying' treatment. There are currently no pest management action thresholds for hemipteran pests of cocoa but guidelines outlined in Table 1 provide a general recommendation). Visual assessment up to hand-height (or two metres above ground level) [25] suggests that the rate of increase does not follow a definite pattern as severely devastated trees can have very low insect numbers. This is contrary to the case for many plant-pest relationships where severely damaged plants tend to have a lot more insect numbers some of which might undoubtedly be difficult to locate. There should, therefore, be a plan for timely treatment or contingency use of insecticide in cocoa farms for mirid and stink bug control. Some of the ways of considering insecticide use in cocoa are depicted in Table 1.

### Table 1. Farm-specific recommendation for mirid and stink bug control in cocoa

| Prediction | Recommended treatment |
| --- | --- |
| 1. Mirid and stink bug damage unlikely (<5 individuals/100 trees | No spraying required; watch out |
| 2. Mirid and stink bug slight/moderate (5-10 individuals/100 trees | Timed spray* spot treatment |
| 3. Mirid and stink bug moderate/severe (>10 individuals/100 trees) | Spray whole farm and repeat approx. 4 weeks later |

*Timed spray – insecticide applied as soon as 3-5% trees or pods are affected. Source: Emmanuel Obeng, CRIG, unpublished data

In planning a pest management programme, it is often necessary to consider the speed of chemical application which comes with a cost. Subsequently, it would be unreasonable to rely on a calendar spray system if large areas are involved. Probably it will be better to use more than one of the programmes suggested above in order to spread the work-load. Nevertheless, farmers are advised to move entirely away from calendar sprays and to optimize the biological control component by spraying strategically. This is difficult in most of Africa where not all farmers have the necessary spraying equipment. In a situation of a second pest, the insect would have to be considered separately or the pest would have to be treated irrespective of the mirid situation as has happened in the past with *Anomis* sp. and other caterpillars.

In June to August in many years, most pests tend to occur intermittently and the economic benefit from spraying is much less obvious than during occurrence of mirids (i.e. September to May). On most cocoa farms a sensible programme should be to aim treatment at the control of mirids and the other important pests such as stink bugs. A two-spray regime programme- two sprays in the first quarter of the year to control mirids and stink bugs attacking cherelles (young cocoa pods) and a further two treatments in the major pest season – would cost

(at 2014 prices) the equivalent of about GH¢80.00 ($26.60) per ha for the insecticide. To this should be added the cost of labour and hired equipment for the treatment. Evidence from revenues from farmers' passbooks suggests that over the past few years most cocoa farms could not repay this cost. However, as mentioned before, farmers who are achieving high yields find it necessary to have some insurance against insect pests and in these cases, the programme outlined would be justified.

The basic plan for insecticide use should, therefore, involve no, two or four strategically timed applications of recommended insecticides and the decision must be based first and foremost on the incidence of the primary pest with the present range of approved insecticides. Additional insecticides should only be used in exceptional cases when the key pest population increases beyond the economic threshold level. The blanket treatment of large areas with insecticides will undoubtedly have ecological consequences that may not be visible initially (unless effective bio-monitoring is carried out), but will manifest later and may be difficult to rectify. Furthermore, there is always the possibility that frequent spraying would encourage the development of insecticide resistant strains of the insect pests, destruction of beneficial non-target arthropods, and environmental pollution (e.g. contamination of water). Negative impacts on birds, reptiles and amphibians, which are of the most efficient predators of insects, are also a very common side effects of the use of insecticides. Moreover, many of the chemicals used to control mirids and other insect pests listed by the WHO in class II or III. Therefore, it is particularly important that they are used only when necessary. Spot spraying (where localized areas are treated, and minimum amount of chemical is used) involves minimal cost and minimum contamination of the environment.

## 6. GOVERNMENT-ASSISTED SPRAYING ('MASS SPRAYING')

Government-assisted spraying of pesticides by recruited spraying teams in cocoa was started in 2001 [26]. This is a government intervention that aims to treat large areas quickly to boost production and increase farmer income. Although the areas under treatment can be overwhelmingly large, many farmers need to plan their own treatment. Under the spraying programme dubbed 'mass spraying', delays in input supply to rural farming communities make timely application of pesticides difficult. Also, there are frequent cases where spraying teams may have been recruited from completely unfamiliar communities and which may hinder or delay spraying activities. Where chemicals are to be applied continuously in large blocks of cocoa farms and especially when these occur on mountain slopes, the provision of marked walking tracks would be an advantage, but the teams are often impatient. Nevertheless, the system appears to be beneficial as more than a 30% annual increase in yield was observed in the 2010/2011 cocoa season. The application of sprays are likely to be much quicker if the farmers made provision for this type of spray application.

A recent survey [9] showed that the spraying of cocoa is partially inefficient due to the large acreages involved, as well as the handling of large amounts of water. Expansion of the size of spraying teams across all communities and quicker filling from water tanks situated in large plantations is suggested as one method of speeding the process. Additionally, a more significant development will be the use of much smaller volumes of water through proper nozzle settings in applying the chemicals. The application technology is one field in which there have been remarkably little developments during the last 10 years. Recently it has been shown that for conventional insecticides, two application of 55 l/ha by motorised mistblower in February to May, and again in September to December at 28 days intervals per season, can give acceptable results on mirids and possibly for stink bugs and other insect pests as well [8]. However, more information is needed on the economics, as well as on the biological efficacy of the insecticides applied under the government spray programme.

## 7. INTEGRATION

When planning an Integrated Pest Management Programme, it is first necessary to consider the need for action regarding each insect pest separately and then to attempt to integrate the control measures into a system which may also include other chemicals such as herbicides, fungicides and fertilizers. Where possible, it is obviously convenient to apply more than one chemical at a time, but this can raise problems with timing and compatibility. The correct timing and frequency of chemical application is necessary for the chemical to be fully effective. It can be very risky in terms of accidental

application of higher rates and also of phytotoxicity to compromise on the timing of a chemical application outside the limits set in the recommendation for its use. Most insecticides need to be applied in relation to a growth stage of the cocoa trees or the development stage of a pest; for many hemipteran pests, the phenology of the host plant is of critical importance. This can make planning easier and it should be possible to devise a system flexible enough to cope with most eventualities provided sufficient time is given to planning.

The use of insecticidal mixtures (cocktails) is becoming more common in cocoa in Ghana [27]. Current cocktail insecticides are mixtures from the two main insecticidal classes of neonicotinoid (e.g. imidacloprid SL 200g/litre, thiamethoxam SC 200 g or 240 g/litre, and acetamiprid EC 100g/litre) mixed with a synthetic pyrethroid (e.g. either bifenthrin EC 100g/litre or deltamethrin EC <10g/litre). Compatibility of chemicals often raises problems, since manufacturers can usually provide information on their own products, but it is difficult to do it for other products. The range of products and formulations, and the possible permutations for their use may make the task for compiling a compatibility table difficult.

## 8. PHEROMONES FOR COCOA INSECT PEST CONTROL

Sex pheromones are widely used for a variety of species, but for mirids, they are not very effective [28]. Aggregation pheromones are being used in some cases, though. Behavioural responses exploited to date in cocoa has been mainly the use of mating and sex attractants which detect adult male mirids only [28] and since the sex ratio of the insect is near 1:1, population levels of the pest can be accurately inferred from trap catches. A great step forward would be made if male produced pheromones could be found and used to monitor female populations directly. Field observation suggests that a great deal of potential exist for use of synthetic pheromones for the cocoa stem borer, *Eulophonotus myrmeleon* Fldr. (Lepidoptera Cossidae) [26].

## 9. CONCLUSION

Strategy for insect pest management in cocoa should be based primarily on environmentally safe pest control alternatives. Intensified research and use of cultural practices and resistant varieties or pheromone-based mating disruption against *Hemiptera*, supplemented with chemical treatments where necessary, will minimize the damaging effects of primary pests. Up to two timely applications of insecticides in February to May or September to December at 28 days intervals may be justified under certain conditions, but further applications should only be made when a serious pest problem arises. Because of the large areas under cocoa in Ghana (≈2 million ha), it is important to take into account the effects of chemicals, including their application, on the environment and on wild life.

Improvements of methods of surveillance and prediction would assist in early and more effective identification of specific pest problems. Development of efficient methods of application would enable a quicker and, therefore, more effective use of chemicals. Decisions on the use of chemicals by smallholder farmers must be based on consideration of factors including application methods, compatibility, correct timing and the resources available.

## ACKNOWLEDGEMENTS

The authors are grateful to Dr E. Owusu-Manu for helpful comments. This paper is published with the kind permission of the Executive Director of the CRIG.

## COMPETING INTERESTS

Authors have declared that no competing interests exist.

## REFERENCES

1. Adu-Acheampong R, Archer S, Leather S. Resistance to dieback disease caused by *Fusarium* and *Lasiodiplodia* species in cacao (*Theobroma cacao* L.) genotypes. Experimental Agriculture. 2012;48:85–98.
2. Entwistle PF. Pests of Cocoa. 1st ed. London: Longman Group Ltd; 1972.
3. Owusu-Manu E. Estimation of cocoa pod losses caused by *Bathycoelia thalassina* (Herrich-Schaeffer) (*Hemiptera*: *Pentatomidae*). Ghana Journal of Agricultural Science. 1976;9:81-83.
4. Crowdy SH. Observations on the pathogenicity of *Calonectria rigidiuscula* (Berk. & Br.) Sacc. on *Theobroma cacao* L. Annals of Applied Biology. 1947;34:45–59.
5. McGhee PS. Biology, ecology and monitoring of the *Pentatomidae* (*Heteroptera*) species complex associated with pome fruit production in Washington. M. S. thesis, Washington State University, Pullman; 1997.
6. McPherson JE, McPherson RM. Stink bugs of economic importance in North of Mexico. CRC, Boca Raton, FL; 2000.
7. Owen H. Further observations on the pathogenicity of *Calonectria rigidiuscula* (Berk & Br.) Sacc. to *Theobroma cacao* L. Annals of Applied Biology. 1956;44:307–321.
8. Owusu-Manu E. Frequency and timing of insecticide application to cocoa mirids. Ghana Journal of Agricultural Sciences. 2001;34:71–76.
9. Adu-Acheampong R, Janice J, van Huis A, Cudjoe AR, Johnson V, Sakyi-Dawson O. et al. The cocoa mirid (*Hemiptera*: *Miridae*) problem: Evidence to support new recommendations on the timing of insecticide application on cocoa in Ghana. International Journal of Tropical Insect Science. 2014;34(1):58–71.
10. Stapley JH, Hammond PS. Large-scale trials with insecticides against capsids on cocoa in Ghana. Empirical Journal of Experimental Agriculture. 1957;27:343.
11. Peterson DG, Telford JN, Bond EF, Prins G. Resistance of cocoa mirids (*Hemiptera*: *Miridae*) in Ghana to lindane and the cyclodiene insecticides and the selection of alternative insecticides. Technical Report, January 15, 1964. Tafo Cocoa Research Institute. 1964;110.
12. Anon. Cocoa resistant to capsids. Rep. Cocoa Res. Inst, Ghana. 1944/45. 1946;27-28.
13. Adu-Acheampong R. Pathogen diversity and host resistance in dieback disease of cocoa caused by *Fusarium decemcellulare* and *Lasiodiplodia theobromae* PhD thesis, Imperial College London. 2009;192.
14. Johnson CG. The relation between capsid numbers, new damage and the state of the canopy, and its significance in the tactics of control. In proceedings of the 3$^{rd}$ International Cocoa Research Conference, 23–29 November 1969, Accra, Ghana. 1971;190–196.
15. Padi B, Owusu GK. Towards an integrated pest management for sustainable cocoa production in Ghana; 2001. Accessed 17 January 2012. Available:http://nationalzoo.si.edu/Conserv

ation and Science/MigratoryBirds/Research/Cacao/padi.cfm

16. Owusu-Manu E. Insecticide residues and tainting in cocoa. Pesticide management and insecticide resistance. Academic Press. INC. New York, San Francisco, London; 1977.

17. Ackonor JB, Abdul-Karimu A. A field study on the susceptibility of cocoa genotypes to infestation by four homopterous insect species. In: Proceedings of the 4th International Cocoa Pests and Diseases Seminar, Accra, Ghana. 2003;112-116:2004.

18. Cocoa Research Institute of Ghana: Cocoa manual – a source book for sustainable cocoa production; 2010.

19. Marchart H, Collingwood CA. Varietal differences in capsid susceptibility. In: Rep. Cocoa Res. Inst. Ghana. 1969-1970. 1972;96-99.

20. Adu-Acheampong R, Padi B, Ackonor JB, Adu-Ampomah Y, Opoku IY. Field performance of some local and international clones of cocoa against infestation by mirids. In proceedings global approaches to cocoa germplasm utilisation and conservation: A final report of the CFC/ICCO/IPGRI project on cocoa germplasm utilisation and conservation: A global approach. 1998-2004. 2007;187-188.

21. Byers RA, Kendall WA, Peaden RN, Viands DW. Field and laboratory selection of *Medicago* plant introductions for resistance to the clover root curculio (*Coleoptera*: *Curculionidae*). Journal of Economic Entomology. 1996;89:1033-1039.

22. Strong DR, Kaya HK, Whipple AV, Child AL, Kraig S, Bondonno M, Dyer K, Maron JL. Entomopathogenic nematodes: Natural enemies of root- feeding caterpillars on bush lupine. Oecologia. 1996;108:167-173.

23. Ackonor JB, Adomako B. Relative abundance of species of homoptera on 30 progenies in Ghana. Journal of the Ghana Science Association. 2004;6:(2):85-89.

24. Williams G. Field observations on the cacao mirids, *Sahlbergella singularis* Hagl. and *Distantiella theobroma* (Dist.), in the gold coast. Bulletin of Entomological Research. 1953;44:101–119. DOI:10.1017/S0007485300022987.

25. Collingwood CA, Marchart H. Chemical control of capsids and other insect pests in cocoa rehabilitation, In proceedings of the 3rd International Cocoa Research Conference. 23–29 November 1969, Accra, Ghana. 1971;89–99.

26. Adu-Acheampong R, Padi B, Ackonor JB. The life cycle of the cocoa stem borer, *Eulophonotus myrmeleon* Fldr. in Ghana. Tropical Science. 2004;44:28-30.

27. Adu-Acheampong R, Ackonor JB. The effect of imidacloprid and mixed pirimiphos-methyl and bifenthrin on non-target arthropods of cocoa. Tropical Science. 2005;45:153–154.

28. Sarfo JE. Behavioural responses of cocoa mirids, *Sahlbergella singularis* Hagl. and *Distantiella theobroma* (Dist.) to sex pheromones PhD thesis, University of Greenwich, UK. 2013;290.

# 2

# Grasscutter (*Thryonomys swinderianus*) Production in West Africa: Prospects, Challenges and Role in Disease Transmission

## L. A. F. Akinola[1*], I. Etela[1] and S. R. Emiero[1]

[1]Department of Animal Science and Fisheries, Faculty of Agriculture University of Port Harcourt, P.M.B 5323 Port Harcourt, Nigeria.

*Authors' contributions*

*This work was carried out in collaboration between all authors. Author LAFA designed the study and drafted the manuscript, Authors IE and SRE edited the manuscript. All authors read and approved the final manuscript.*

Editor(s):
(1) Hugo Daniel Solana, Department of Biological Science, National University of Central Buenos Aires, Argentina.
(2) Zhen-Yu Du, School of Life Science, East China Normal University, China.
Reviewers:
(1) Anonymous, Ghana.
(2) Godwin S. Ikani Wogar, Department of Animal Science, University of Calabar, Nigeria.

## ABSTRACT

This review brings together and consolidates the various researches that had been undertaken in grass-cutter with the aim of providing adequate information that will be capable of improving and sustaining the production of the animal as well as its consumption in West Africa. Given the above scenario, this paper reviewed the results obtained by different researchers on feeding and nutritional requirement of the grasscutter, housing, performance, anatomy and morphology, some environmental issues, the challenges and the role of grasscutter in disease transmission. It was clear from this study that grasscutter is widely acceptable, utilizes inexpensive feed to produce good meat of high biological value, survives in simple housing apartments when confined, has good litter size and short generation interval, has simple anatomical dispositions which helps in breeding and they are capable of adapting to intensive rearing environments. However, some challenges of the grasscutter production include irregular supply of breeding stock, environmental issues, poor processing and marketing plan, lack of balanced diet, poor producer training and education, inadequate infrastructural development, poor information dissemination, incidences of diseases and

*Corresponding author: E-mail: letorn.akinola@uniport.edu.ng, letorn_akinola@yahoo.com;*

mortality. With good understanding of the prospects and positive utilization of the challenges, there will be great improvement and sustainability in the production of the animal, such that more jobs will be created, more income generated and protein intake in the sub-region will be increased. This overview therefore highlights essential elements for sustainable grasscutter production and consumption in West Africa as well as suggests the direction of further research on grasscutter production.

Keywords: Grasscutter; feeding; housing; performance; diseases transmission; challenges.

## 1. INTRODUCTION

Wildlife has great potentials for meat production and serves as an important source of the highly desired animal protein for both in urban and rural communities in Africa [1]. In Nigeria there is an abundant variety of wildlife resources capable of supporting the protein intake of the populace. But in recent times, there had been significant short fall between the production and supply of animal protein to feed this ever increasing population [2]. To arrest this unacceptable trend, efforts had been directed towards boosting the micro-livestock sector. Among the micro-livestock species is the grasscutter or cane rat (*Thryonomys swinderianus)* popularly called Oya by the Yorubas, Nchi by the Igbos and Gegbi in Hausa Language in Nigeria, Akranti/Akrantie in Ghana and simply grasscutter in other West African Countries (Figs. 1 and 2).

Grasscutter is a hystricomorphic rodents widely distributed in the African sub-region and exploited in most areas as a source of animal protein [3]. It is a heavily built animal with round muzzle, small round ears, short tail and hursh bristly fur [4]. Apart from being the most preferred [5,6], it is the most expensive meat in most countries in West Africa including Nigeria, Togo, Benin, Ghana and Cote d' Voire [7]. It contributes to both local and foreign earnings in some of these countries [8]. Most rural populations in Nigeria depend on bush meat for their dietary protein supply [9]. Grasscutter meat has been reported to have higher nutritional value [10] and meat yield [11]. Consequently, the demand for livestock products could be solved through the production of grasscutter meat [12]. It is considered a delicacy in West and Central Africa [3]. It serves as a steady source of animal protein in many rural areas of Nigeria and other West Africa countries like Benin, Ghana, Togo, Cot D' Voire [13] and South Africa [14]. Most Chinese who are resident in Nigeria cherish grass-cutter meat as regular meal and form a delicacy for entertainment for their guest from abroad [15]. This preference for grasscutter is attributable to its high carcass quality and protein that is comparable to that of poultry, especially turkey and other domesticated livestock like rabbit, cattle, sheep, goat and pig [16,17]. A crude protein content of 22.7% had been reported for grasscutter meat. This value is higher than the crude protein values of 20.7% for rabbit meat, 19% for chicken meat and 18.2% for beef and 22.2% for turkey [18]. Its mature weight has been reported to be more than 9kg for males and 5-7 kg for females [17]. It is therefore obvious that with the ever increasing human population and obvious protein shortage in Africa, there is need to explore other means of providing readily and acceptable meat on short term basis. A good understanding of the principles and techniques of production will lead to profitable grasscutter business [19]. Thus with appropriate information regarding the prospects and possible challenges of grasscutter production, more farmers will engage in the production of grass-cutter. This will go a long way in alleviating poverty, reducing hunger, creating jobs, improving income and the immune system of Nigerians against diseases associated with low level of protein intake.

## 2. PROSPECTS OF GRASSCUTTER PRODUCTION

Research has been carried out on the nutritional requirement of grass-cutter [20,21], reproductive performance [22-24], housing [24,25], environmental and reproductive biology [26] disease and mortality [27,28].

### 2.1 Feeding and Nutritional requirement of Grasscutter

The grasscutter is primarily a herbivore, but in confinement, they require supplementary feed [29]. It is easy to feed and it is a good food transformer and practices coprophagy [30]. The large caecum which forms part of the digestive tract of the grasscutter is adapted predominantly to microbial digestion of feeds [31]. Grasscutter prefers mainly thick-stemmed grass species [32]. The feeding habits of grasscutter and other rodents (e.g. rabbits) are directly opposite. Whereas the grasscutter prefers to eat stalks to leaves, the rabbit, for example on the contrary choose the leaves and waste stem [32,33]. This habit leads to waste of feed resources by the grasscutter, especially during the dry season when there is scarcity of grass. Thus, irrespective of the kind of forage, grasscutters first eat stalks, the bark of twigs and finally some leaves [20]. This eating habit causes wearing of the animal's teeth, and unfortunately leads to high forage wastage [34].

They prefer grasses with lots of moisture and soluble carbohydrate [35,36], preferring succulent grasses and stems like sugar cane [37,38] and Guatamala plants [39]. Grasscutter can also be raised by feeding them with kitchen left-overs [40]. They also eat fallen fruits, nuts and many kinds of cultivated crops [41]. Some grasses and plants that are highly utilized by grasscutter during the dry and wet season have also been identified [42]. These include: *Pennisetum purpureum* (elephant grass), *Saccharum officianarium* (sugar cane), *Zea mays* (maize), *Sorghum vulgare* (Guinea corn), *Oryza sativa* (rice), *Andropogon gayanus* (gamba grass) etc. According to [20] grasscutters show high preference for grass and particularly favor *Pennisetum purpureum* (elephant grass) and *Panicum maximum*. Good carcass quality and highest live weight was reported by [21] when grasscutters were fed 2000 KcalMEkg⁻ in combination with chopped elephant grass (*Pennisetum purpureum*). Feeds containing 12-20% crude protein have been reported to be suitable for grasscutter [43] while 18% crude protein (CP) was required for optimum growth of grasscutter from weaning to reproductive stage [44]. According to [45] gestating grasscutter give optimum result in terms of litter size, birth weight of pups and feed conversion ratio when 14% CP is included in their diet. Thus, [46] recommended

that the optimum energy requirement of growing grasscutter is 2200 KcalME/kg in the humid tropics when the CP is 18% while [47] stated that the preferred source of fibre for the growing grasscutter is palm kernel cake.

Table 1 shows the nutrient requirement of grasscutter as reported by [48,49] while Table 2 gives the quantity of feed consumed by grasscutter per day at different ages as stated by [17].

According to [30,50] the water intake of the grasscutter is reduced when the temperature is hot and more when the out-door temperature is low. They recommended this as a subject for further research since the reason for this unexpected behaviour was not yet understood.

## 2.2 Production System/ Housing

Production system of grasscutter can be classified according to production objectives into commercial or subsistence [51]. Report has shown that the animal can be bred and kept in boxes, empty drums, Poly Vinyl Chloride (PVC) pipes and enclosures among the rural communities and even in some urban areas among people with adequate space [52]. The report by [24] showed that cane rat litters reared for six weeks after parturition in block-cement pens had comparative advantage in terms of weight gain and the rate of survival compared to those reared in iron cages. [25] therefore advocated for use of block pens (with wood shavings on the concrete floor) at the beginning of rearing grasscutter as they recorded lesser deaths of the animal than those in iron cages. The housing of grasscutter consist of stables and pens equipped with cages and hutches made of good materials and blocks, strong enough to stop the very sharp incisors of the animal [17,53].

**Fig. 1. A family of grass-cutter**

**Fig. 2. A mature Grass-cutter**

## 2.3 Dentition in the Grasscutter

The dentition of the grassscutter is typical of the rodentia with 10 pairs of teeth [17,51,54]. These comprise of one incisor (1I), no canine (0C), one premolar (1P) and three molars (3M). [55] reported that the incisors of grass-cutter are probably the most powerfully built of any African rodent. The time of eruption of the teeth helps the farmer to know the appropriate type of feeding to be offered to the animal and know the age of the animal. Table 3 shows a summary of a 51 weeks study on the eruption of grass-cutter teeth as stated by [56].

## 2.4 Reproductive Performance of Grasscutter

The reproductive outputs are measured according to animal maturity, litter size, length of inter-birth interval and age at last reproduction [23]. Thus, [24] identified some reproductive parameters of breeding grass-cutters to include having signs of mating on the 3rd day and 7th day, gestation interval of 155±8 days and 157±3 days, average litter size of 4 and 5, sex ratio of litters, 3 males : 1 female and 3 males: 2 females and mean weight gain of litter at six weeks of weaning 539±12g and 595±12g respectively for grass-cutter housed in iron cage (IC) and block–cement pen (BP) having same size of 120cm x 75cm x 30cm. Studies on the reproductive performance of female grass-cutter (does) at first parity in the humid tropical environment showed that 50% of the does studied had open vaginal status at first paring while the remaining 50% were either closed or plugged [57]. The result also indicated that 50% of the does conceived at second exposure while 18.8% conceived at first exposure with more female off-springs. This confirmed the finding by [40,58] that grass-cutter and rabbits show variations in reproductive activity even though they are identified as induced ovulators. Studies by [59] revealed that the grasscutter has a mean gestation length of 163.11±1.58 days (with a range of 152-170 days), litter size of 4 (with a range of 2-7), mean birth weight of 117.70±34.08g (with male birth weight being generally heavier, 118.10±27.70g than females, 100.90±27.50g). They observed that breeding occurred in January, March, April, June, July, November and December with 67% of the parturition occurring at night. Further studies by the authors showed that litter weight decreased with increase in litter size, but did not influence the growth performance of the baby grasscutter during the first months of life. A mean litter size of 4 had been previously reported by [60].

## 2.5 Anatomy and Morphology of Grass-cutter

A good knowledge of the anatomical dispositions of the reproductive organs of grass-cutter is vital to the understanding of the reproductive biology of animals and provides information which would assist in the breeding of the cane rat and improve its domestication. The male reproductive organ of the grasscutter is similar to that reported by previous authors [61]. The testes of the cane rat has ovoid shape with creamy white coloration covered with stroma [62,63], typical of rodents. The surface of the testes of the cane rat showed the presence of *Tunica vaginalis propria* with radical septa (*Septuli testes*) of pyramidal shape [64]. The shape of the epididymis is sigmoid unlike that of a typical rodent and the distal part of the caudal epididymis is convoluted as in mammals [63] and provides useful information in the comparative regional anatomy of rodents. The morphology and morphometry of the grasscutter male accessory sex gland has also been reported [65]. According to [29] the best way to identify the different sexes is by studying the distance between the anus and the genital organs which is usually wider in the males and almost double that obtainable in the female.

## 2.6 Environmental Issues in Grass-cutter Production

The interaction of the grasscutter in captivity with its immediate environment appears very important in relation to mortality, improve reproductive competence, improve health and overall performance. [26] reported that a pit pen housing design was able to maintain the room temperature of the house and pens (24.43 – 30.71°C) against the diurnal fluctuations outside the building (25.86 – 34.71°C). They however stated that the relative humidity inside the building (67.57 – 85.80%) showed a tendency to fluctuate with the ambient relative humidity outside the building. They concluded that if captive grasscutter were housed in environment similar to the night period, they are likely to shed their nocturnal habit and be more active during the daytime, with the likelihood of increased productivity.

## 2.7 Other Benefits of the Grasscutter

Despite a lack of defined or measurable indications for its contribution to the gross domestic product (GDP), the grasscutter subsector has been recognized as an important economic tool for rural poverty alleviation and household food and nutrition security [66,67]. The grasscutter is a considerable income earner for both the small scale peri-urban or rural livestock producer in the country. It also contributes to both local and export earning of countries like Kenya, Benin republic and Nigeria [8]. It is known to be economically important as an agricultural pest and its meat is widely accepted by all classes of people. The smoked grasscutter could serve as a source of foreign earnings when it is well packaged and exported.

Grasscutter meat is cheaper to produce than most other traditional livestock and its meat is more valuable and appreciated by local population. It has been shown that with only five mature grasscutters (4 females and 1 male), a household is nutritionally secured for 6 months to one year [68]. In times of droughts and related calamities, grasscutter serves as a critical source of animal protein.

During important occasions and ceremonies, grasscutters are heavily consumed by many households in rural and urban areas in Nigeria. Dried grasscutter meat is used to serve the elders during traditional rights like marriages, excursion and Chieftaincy installations. This confirms that grasscutter is the favorite bush meat species. The hair or fur is used to make decoration and the teeth are used to perform traditional card reading in place of cowries especially in the Southern part of the country.

### Table 1. Nutrient requirement of an adult grass-cutter

| Component | % dry matter basis |
|---|---|
| Crude protein | 12 to 18.5 |
| Crude lipid | 2.5 to 4.5 |
| Crude fibre | 25 to 45 |
| Ash | 2.5 to 4.5 |
| Nitrogen free extract | 45 to 65 |
| Neutral detergent fibre | 42 to 64 |
| Acid detergent fibre | 25 to 35 |
| Acid detergent lignin | 3 to 8 |

Source: Mensah [48,49] (1995,2005)

### Table 2. Quantity of feed consumed by grass-cutter per day

| Subject | Grass/forage (g) | Supplement (g) |
|---|---|---|
| Cutling (Young grass-cutter) | 10 – 150 | 10 -50 |
| Weaner/grower grass-cutter | 152 – 250 | 51 – 100 |
| Adult grass-cutter | 251 – 450 | 101 – 200 |

Source: Fayenuwo et al. (2003) [17]

### Table 3. Cutting of teeth (teeth eruption) periods in grass-cutter

| Age/Period | Teething per half jaw | | | | |
|---|---|---|---|---|---|
| | Incisor (I) | Canine (C) | Premolar (P) | Molar (M) | No. of teeth (%) |
| Birth | $P1^1$(n=51) | - | $Op^1$ (n=51) | - | 40 |
| 2 weeks | ,, | - | $Ap^1$ (n=48) | $Om^1$ (n=48) | 60 |
| 2-4 months | ,, | - | $Pp^1$ (n=42) | $Pm^1$ (n=42) | 60 |
| 5-8 months | ,, | - | ,, | $Pm^2$ (n=35) | 80 |
| 9 months | ,, | - | ,, | $Pm^3$ (n=33) | 100 |

Source: Fayenuwo et al. 2005 [56]
= Tooth present = Tooth absent

$P1^1$ = Presence of one incisor          $Op^1$ = Opening of premolar
$Ap^1$ = Appearance of the premolar          $Pp^1$ = Presence of premolar
$Om^1$ = Opening of the $1^{st}$ molar          $Pm^1$ = Presence of $1^{st}$ molar
$Pm^2$ = Presence of the $2^{nd}$ molar          $Pm^3$ = Presence of $3^{rd}$ molar
Dental formular:          I =Incisor, C = Canine
I(1) C(0) P(1) M(3) = 20          P = Premolar, M = Molar

## 3. CHALLENGES OF GRASS-CUTTER PRODUCTION

According to [69], some of the major problems encountered by grass-cutter farmers include: high initial capital, stock procurement, time constraint, inadequate medical attendant, disagreement with landlords over space to rear grass-cutter and inadequate follow-up by extension services. Recently, [25] ranked some constraints encountered by the grass-cutter farmer from the most severe to the least as follows: lack of capital, insufficient feed, disease, housing, lack of enough land, handling and lack of knowledge of rearing grass-cutter. Other challenges of grass-cutter include irregular stock supply, environmental issues, processing and marketing, feeding, producer's training and education, infrastructure development, poor information dissemination and disease/mortality.

### 3.1 Irregular Supply

The production of grass-cutter is a relatively novel practice. Although most breeding stock and cane rat meat is still obtained by hunting and trapping of the animals which does not ensure steady and regular supply of the meat [70] as well as the breeding stock.

### 3.2 Environmental Issues

The collection of grass-cutter from the wild is attended by the destruction of the environment through setting of bush fires by hunters [3,8,66]. This leads to the destruction of valuable plants, animal life and tampers with the ecosystem [70]. Thus, there is need to domesticate the animals in order to avoid the problems associated with bush burning.

### 3.3 Poor Processing and Marketing Plan

Most small scale and medium-scale farmers do not provide a good plan for processing and marketing of their grass-cutter at the initial stage of their business, as a result when the animal attain market weight, only a few buyers are seen. The farmer then devices a means of marketing (live or processed) grass-cutters while operating at a loss or reduced profit because of the extension in feeding time. This leads to problem associated with storage facilities, waste disposal, and disease contamination, accompanied by reduction in selling price.

### 3.4 Lack of Balanced Diet

The domestication of cane rat has its own teething constrains, which include the need to provide regular supply of feeds rich and balanced in nutrients [70]. It has been observed that grasscutters prefer grasses such as elephant grass, sugar cane, guinea grass with succulent stalk [17] which may not be readily available. Furthermore, grasscutter reared in captivity on forages and grasses alone does not do well compared to those living in the wild. This is because grasscutter normally obtains balanced nutrient from a variety of feeds such as forages tubers, grains, nuts, herb etc in their natural habitat or in the wild.

### 3.5 Producer Training and Education

The education of farmers has been found to be one of the major factors affecting adoption of new technologies [71]. Intensified education on grasscutter breeding and production to save the animal from extinction will reduce poverty and create employment. Report by [72] showed that most grasscutter farmers (90%) in Oyo State, Nigeria, had completed one form of formal education or another, implying that education is a variable which widens the mental horizon and predisposes farmers to new ideas. This results in having better access to knowledge and information that will be beneficial to the production and management of grascutter.

### 3.6 Infrastructure Development

Grasscutter production is mainly concentrated in the rural areas which are characterized by poor infrastructural facilities such as road and telecommunication network. Installation of these facilities would open these areas for development [73] and enhance access by the producers to market for purchase of inputs and sales of his products. Construction of good roads would help the extension services providers to reach as many producers as possible allowing training on new production technologies which will result in increased productivity of grasscutter.

### 3.7 Poor Information Dissemination

The grasscutter farmers in Nigeria as in other developing countries are faced with poor information dissemination about the challenges (such as disease out breaks, feeding, breeding, housing, marketing and lack of credit facilities) facing their production. Information is an essential ingredient in agricultural development programmes but Nigerian farmers seldom feel the impact of agricultural innovations either because they have no access to such vital information or because it is poorly disseminated [74]. The extent of information needed by grasscutter producers had been reported [72]. They identified grasscutter diseases, housing pattern and equipment needed for production as the 1st, 2nd and 3rd most essential information needed by the grasscutter farmer. Table 4 gives their findings and rating of the information needs of grasscutter farmer. These problems can be solved through well-designed and implemented information dissemination and awareness programmes, including seminar which will endow all the stakeholders in the sector with necessary knowledge and skill [75]. Thus, there is need for networking amongst grasscutter farmers to enable them acquire and share knowledge, views and experiences among themselves and with all other stakeholders along the grasscutter value chain.

### 3.8 Mortality and Disease of Grasscutter

Disease is one of the most important limiting factors to profit in many livestock enterprises in the tropics [76]. Apart from inadequate and unbalanced feeding, high disease prevalence and associated high neonatal mortality constitute a major obstacle to the promotion of large scale holding of livestock [77]. It is important to note that the severity of diseases depends on the nutritional state of the animals, especially during the dry season when feed is inadequate in quantity and quality [78]. Also, the incidence, severity and prevalence of disease had been shown to vary with the management system [67].

According to [26] more grasscutters died when they were housed in iron cages at the beginning of farm operation than when they were housed in block-cement pens. In another research, [28] observed that more grasscutters (28) died of pneumonia among those kept in concrete cages with cemented floor while only (3) died among those kept in iron cages as a result of trauma and dystocia (difficulty in parturition). Outbreak of intestinal coccidiosis was observed in cane rat [27] while [79] identified twenty major disease/disease conditions affecting grass-cutter in captivity. Gastro-intestinal disorders, caused by helminthes parasite had also been identified in the grasscutter [80]. Reports by [81] showed that grasscutters can be infected with trypanosomes, although without obvious clinical disease.

Preliminary studies by [80] on the captive grasscutter in Cameroon showed the occurrence of ectoparasite such as fleas (Xenopsylla sp) and endoparasite like cestode (Hymenolopsis sp) and nematode (Heterakis sp). In another work by [82] in Ghana, four species of tick namely Rhipicephalus simpsoni, Ixodes aulacodi, Ixodes sp and Haemaphysalis parmata, six species of helminthes comprising of 2 species of cestodes (Furhmanella transvalensis, Railettina mahone) and 4 species of nematodes (Longistriata spira, Trachyphanyx natalensis, Paralibyostongylus vondwei and Trichuris paravispicularis) were also found.

## 4. ROLE IN DISEASE TRANSMISSION

It is interesting to note that the grasscutter had not been traced to harbor pathogens that can affect humans. Ebola virus disease for example, had been traced to chimpanzees, gorillas and bonobos and currently spreading to humans when the meat is handled or consumed. Also, gorillas and some other apes may also carry other diseases as simian foamy virus, smallpox, chicken pox, tuberculosis, measles, rubella, yellow fever and yaws (http://en.wikipedia.org/wiki/Bushmeat). It has occurred, on numerous occasions, that people who ate apes have caught such diseases or even died [83]. Thus, apart from posing a significant risk to the people who eat the meat, it poses great risk to the human population as a whole, as it opens a doorway through which animal viruses can be transmitted to humans. Other bush meat like the African squirrels (Heliosciurus, Funisciurus) have also been implicated as reservoirs of the monkey pox virus in the Democratic Republic of Congo (http//:enwikipedia.org/wiki/Bushmeat), implying that the use of their meat may serve as a means of transmission of these viruses to humans. According to [84] research in Africa has proven that Ebola disease can only occur through the handling of infested chimpanzes, gorillas, fruit bats, monkeys, forest antelopes and porcupines found dead or ill in the rainforest, so Ghanians can continue to enjoy their bushmeat delicacies, provided it is handled safely and prepared without any contamination. The commercial production of grasscutter will therefore be of great benefit to lovers of bush meat since the animal has not been linked to pathogens.

**Table 4. Information needs of farmers in grass-cutter production**

| S/N | Information Needs | Scores |
|-----|-------------------|--------|
| 1 | Rabbit housing pattern | 2nd |
| 2 | Cleaning of housing unit | 7th |
| 3 | Sources of stable grass-cutter breed | 9th |
| 4 | Types of feed available | 12th |
| 5 | Weaning | 8th |
| 6 | Equipment required for grass-cutter production | 3rd |
| 7 | Appropriate number of grass-cutter required in cages | 15th |
| 8 | Incentive on grass-cutter | 12th |
| 9 | Identifying various grass-cutter disease | 1st |
| 10 | Selection of foundation stock | 11th |
| 11 | Marketing of grass-cutter | 5th |
| 12 | Record keeping | 4th |
| 13 | Control of pests and diseases of grass-cutter | 6th |
| 14 | Method of mating | 10th |
| 15 | Ovulation and heat period | 14th |

*Source: Fakoya et al. (2008) [72]*

## 5. CONCLUSION

This study has showed that the grasscutter, a wild African rodent can be successfully domesticated as some of the essential elements in the successful production of the animal were reviewed. The feeding and nutritional requirement, production system and housing, dentition, reproductive performance, anatomy and morphology, environmental issues and the benefits have all been studied. Detailed information on the challenges, including irregular supply of the stock animals, environmental issues, feeding, poor producer training and education, infrastructural development, poor information dissemination, mortality and diseases among others have also been reviewed. This animal which provides juicy and palatable meat, and is highly preferred in meals of both Nigerians and foreigners, without any fear of disease transmission, therefore, offers suitable opportunities for enhancing livelihood and revenue generation in rural and urban areas in the sub-region Consequently, this review creates more opportunity for the grass-cutter farmers and intending farmers to easily sustain their businesses, create more jobs, increase income and increase protein consumption of the growing populace while it serves to assure the consumers of bushmeat (commercial grasscutter in this case) of the safety of the meat.

## COMPETING INTERESTS

Authors have declared that no competing interests exist.

## REFERENCES

1. Fonweban JN, Njwe RM. Feed utilization of life weight gain by the Africa giant rat (*Cricetonys gambianus*, water house) at Dischana in Cameroon. Tropiculture. 1990;8(3):118-120.

2. Akpan IA, Wogar GSI, Effiong OO, Akpanenua EJ. Growth performance of grass-cutter fed diets treated with urea and urine solution. Proc. of 34th Ann. Conf. of Nig. Soc. for Anim. Uyo, Nigeria. 2009;163-164.

3. National Research Council. Microlivestock: Little known small animals with a promising economic future, National Academy Press, Washington DC, USA; 1991.

4. Taiwo OW. Wealth creation through commercial grass-cutter farming. Production Techniques for Good Performance. 2006;67.

5. Clottey JA. Relation of physical body composition to meat yield in the grass-cutter (*Thryonomys swinderianus* Temminck). Ghana Journal of Science. 1981;21(12):110-115.

6. Martin GH. West Africa: Carcass composition and palatability of some animal commonly used as food world animal review. 1985;(53):40-44.

7. Asibey EOA, Addo PG. The grass-cutter, a poising animal for meat production, in: African perspective, practices and polices supporting sustainable development (Turnham, D. Ed) scandinavan Seminar College, Denmok; in Association with weaver press Harare. Zimbabwe; 2000. Available: www.cdr.dk/sscafrica/asad-gh.htm

8. Ntiamoa-Baidu V. Sustainable use of bush meat. Wildlife Devt. Plan. 1998-2003; 1998.

9. Food and Agricultural Organization of the United Nations. Integrated crops and livestock in West Africa. Animal Production and Health Paper 41. FAO, Rome Italy; 1983.

10. Opara MN. The grasscutter 1: A livestock of tomorrow. Res. J. Forestry. 2010;4:119-135.

11. Omole AJ, Ayodeji IO, Ashaye OA, Tiamiyu AK. Effect of scalding and flaming

methods of processing on physiochemical and organoleptic properties of grasscutter meat. J. Applied Sci. Res. 2005;1:249-252.

12. Adekola AG, Ogunsola DS. Determinants of productivity level of commercial grasscutter farming in Oyo state. Proc. Int'l. Conf. on Global Food Crisis, 19th-24th April, Owerri, Nigeria. 2009;15-21.

13. Ogunsanmi AO, Ozegbe PC, Ogunjobi O, Taiwo VO, Adu JO. Haematological, plasma biochemistry and whole blood minerals of the captive adult African grasscutter (*T. swinderianus* Temminck). Trop. Vet. 2002;20:27

14. Van-Zyl A, Van der Merwe M, Blignaut AS. Meat quality and carcass characteristics of the Vondo, *Thryonomys swinderianus*. South African Journal of Animal Science. 1999;29:120-121.

15. The Thy Consulting (2006). How to become your own boss through high profitable Agricultural businesses. Grasscutter farming. 2006;11. Available:http://grasscutterfarming.tripod.com

16. Ajayi SS, Tewe OO. Food preference and carcass composition of grass-cutter (*Throyonomis swinderianus*) in captive. African. J. Eco. 1980;18:13-14.

17. Fayenuwo JO, Akande M, Taiwo AA, Adebayo AO, Saka JO, Lawal BO, Tiamiyu AK, Oyekan PO. Guidelines for grasscutter rearing. Technical Bulletin. IAR&T Ibadan Nigeria. 2003;38.

18. Anieunam AS. The place of bush meat in the supply of animal protein in Nigeria. Seminar Paper, Department of Animal Science, University of Calabar, Nigeria; 2005.

19. Akinola LAF. Grasscutter farming: A new initiative in protein supply. An Invited paper presented at Agricultural Product Development Strategy Workshop organized by Rivers State Sustainable Development Agency (RSSDA), held on 9-10th Sept. 2008 at The Elkan Terrace, Abacha Road, Port Harcourt, Rivers State, Nigeria; 2008.

20. Mensah GA, Okeyo AM. Continued harvest of the divers African animal genetic resources from the wild through domestication as a strategy for sustainable use: A case of the larger grass-cutter (*Thryonomys swinderianus)*; 2005. Available: http//agtr.ilri.cigar.org/Case-study/Mensah/Mensah.htm

21. Henry AJ, Njume GN. Effect of varied energy levels on the carcass characteristics of grasscutter (*Thryonomys swinderianus*). Proc. Of 33rd Ann. Conf. of the Nig. Soc. for Anim. Prod. Ayetoro, Ogun State, Nigeria. 2008;168-170.

22. Heath E, Olusanya S. Anatomy and physiology of tropical livestock. Longman Scientific and Technical. 1985;138.

23. Redford KH, Godshalk R, Asher K. What about the wild animals (mammals) of some economic importance: Problems and Prospects. Proc. of the National Forestry Workshop on Strategies for the management of endangered environment and species. R.D. Adegbehin and R.D. Gbadegesin eds. 1995;161-178.

24. Ogunjobi JA. Reproductive performance of cane rat (*Thryonomys swinderianus* Temminck 1827) breeding stocks and littesr rate of survival reared using two common housing materials. Proc. of 33rd Ann. Conf. of Nig. Soc. for Anim. Prod. Ayetoro, Ogun State, Nigeria. 2008;208-210.

25. Ogunjobi JA, Inah EIs. Seasonal mortality among farmed cane rat reared using two different housing materials. Proc. of 33rd Ann. Conf. of Nig. Soc. for Anim. Prod. Ayetoro, Ogun State, Nigeria. 2008;136-138.

26. Williams OS, Ola SI, Boladuro BA, Badmus RT. Diurnal variation in ambient temperature and humidity in a pit pen grass-cutter (*Thryonomys swinderianus*) house in Ile- Ife. Proc. 36trh Ann. Conf. of Nig. Soc. For Anim. Prod. held at Univ. of Auja, Nigeria. 2011;111-113.

27. Kasali OB, Majaro OM, Dipeolu OO. An outbreak of intestinal coccidiosis in a colony of can rats in Nigeria. Nig. J. Vet. Med. 1982;9(2):3-5.

28. Fatokun BO, Ogunjobi JO, Olajide OB, Ukandu P. Clinical investigations on the causes of mortrality in grasscutter (*Thryonomys swinderianus*, T. 1827) held in captive within Ibadan metropolis in Oyo State, Nigeria. Proc. of 35th Ann. Conf. of Nig. Soc. for Anim. Prod. held at Univ. of Ibadan, Nigeria. 2010;751-753.

29. Ayodele IA, Meduna AJ. Essentials of Grasscutter Farming. Hope Publications, Ibadan, Nigeria. Pp. 37 BBC News. Scientist find new strain of HIV. 2 August 2009; 2007.

30. Holzer R. Mensah GA, Baptist R. Practical aspect of grass-cutter (*Thryonomys*

*swinderianus*) breeding, particulars of coprophagy. Rev. Elev. Med-vet. Pays. Trop. 1986;39(2):247-252.

31. Alaogninouwa T, Agba KC, Agossou E, Kpodekon M. Anatomical, histological and functional specificities of the digestive tract in the male grasscutter (*Thryonomys swinderianus* Temminck*).* Anatomy, Histology, Embryol. 1996;25:15-21.

32. Schrage R, Yewandan LT. In: Raising Grass-cutters Dentsche Gesellschaft for technische Zusamimenarbeit (GTZ) Gmbh, Eschborn, Germany. 1999;99.

33. Vietmeyer ND. Board on Science and Technology for International Development National Research Council, National Academic Press. Washington DC. 1991;233-240.

34. Adu EK. Constraints to grass-cutter production in Ghana. Proc. of the Int'l forum on grass-cutter. Institute of Local Government Studies. Accra. Ghana, T. Antoh, R Weidinger, J Ahiaba, A Carrilo (eds). Ministry of Food and Agriculture. Accra. Ghana. 2005;44-50.

35. Onadeko SA. The reproductive ecology of the grasscutter (*Thryonomys swinderianus*) in captivity. Ph.D Thesis, Dept. of Wildlife and Fisheries Management, Univ. of Ibadan; 1996.

36. Agbelusi E. Ranching grasscutter (*Thryonomys swinderianus* Temminck) for meat production in the humid forest zone of Nigeria. FAO Jul Publication; 1997.

37. Adu JA. Manual of grasscutter husbandry. Adagro Lagos. 1995;25-30.

38. Awah AA. Introduction to minilivestock development as a sustainable agricultural business. Workshop Paper presented at Agric. Dev. Prog. (ADP) in Anambra, Imo, Rivers and Cross Rivers States; 2000.

39. Ndi MJ. Cane rat farming manual. Cameroun Youth Dev. Centre. Promotion for grasscutter production. No. 3 Series. 2004;7-200.

40. Addo PG. Domesticating the wild grasscutter (*Thryonomys swinderianus* Temminck 1827) under laboratory conditions. Ph.D Thesis, Univ. of Ghana, Legon, Ghana; 1997.

41. Fitzinger F. Can Rats. In: Walker's mammals of the world, Nowak, R (Ed). The Hopkins Univ. Press, Maryland. 1995;1650-1651.

42. Ebenebe CI. Identification and analysis of some plants utilized bt the grasscutter (*Thryonomys swinderianus*

43. Temminck) at Akpaka forest reserve, Onitsha, Nigeria. Proc. of 10th Ann. Conf. of Anim. Sci. Assos. of Nig (ASAN). Univ. of Ado-Ekiti, Nigeria. 2005;251-254.

43. Meduna AJ. Preliminary observation on cane rat (*Thryonomys swinderianus*, Temminck) feeding and breeding. Proc. of the 27th Ann. Conf of the Nig. Soc. For Anim. Prod. Federal College of Tech., Akure. 2002;304-305.

44. Kusi C, Tuah AK, Annor SY, Djang-Fordjour KT. Determination of dietary crude protein level required for optimum growth of the grasscutter in captivity. Livestock Res. for Rural Devt. 2012;24(10). Available: www Irrd.cipav.org.co/Irrd24/10/kusi24176.htm

45. Wogar GSI. Performance of gestating grasscutter (*Thryonomys swinderianus*) fed cassava-based diets with graded protein levels. Asain J. of Anim. Sci. 2011;5:373-380.

46. Wogar GSI, Effiong OO, Nsa EE. Performance of growing grasscutter (*Thryonomys swinderianus*) fed diets with graded energy levels Journal of Agric., Biotechnology and Ecology. 2011;4(3):134-139.

47. Wogar GSI. Performance of growing grasscutter on different fibre sources. Pakistan Journal of Nutrition. 2012;11(1):51-53.

48. Mensah GA. *Futteraufnahme und Verdaulichkeit beim Grasnager (Thryonomys swinderianus)*. Thèse de Doctorat, Institut 480, Université de Hohenheim, Allemagne. 1993;107.

49. Mensah GA. Consommation et digestibilité alimentaire chez l'aulacode *Thryonomys swinderianus*. *Tropicultura*. 1995;13(3): 123-124.

50. Mensah GA. General presentation of the breeding of grass-cutter, history and distribution in Africa. Proc. Of Int'l Seminar on Intensive Breeding of Wildlife Animals for the Purpose of Food in Liberville, Gabon Project, DGEG/VSF/ADIE/CARPE/UE. 2000;45-59.

51. Olomu JM, Ezieshi VE, Orheruata AM. Grasscutter Production in Nigeria. Principles and Practice. Jachem Publishers. 2003;62.

52. Adu EK, Wallace PA, Ocloo TO. Efficacy of sex determination in the greater cane rate, *Thryonomous swinderianus,* Temminck. Trop. J. Anim. Prod. 2002;34(1):27-33.

53. Lameed GSA. Grasscutter/ Canerat (*Thryonomys swinderianus)* Farming in the Tropics: The possibilities and Prospects for would-be Farmers. Peerless Prints, Nigeria. 2008;14-21.

54. Schrage R, Yewaden LT. Abrege D'aulacodiculture. Deutsche Gesellschaft for Technische Zusammen Anbait (GTZ) Gmbtt. 1995;103.

55. Fayenuwo JO, Akande M, Ogundola FI, Oluokun JA. Moor J. Agric. Res. 2001;(2):141-146.

56. Fayenuwo JO, Taiwo AA, Oluokun JA, Akande M, Adebowale EA. Observations on teeth development in captive-bred grasscutter (*Thryonomys swinderianus* Temminck). Proc. of 10th Ann. Conf. of Anim. Sci. Assos. of Nig. Univ. of Ado-Ekiti, Nigeria. 2005;248-250.

57. Henry AJ. Reproductive performance of grasscutter does at first parity reared in humid tropical environment. Proc. of 35th Ann. Conf. of Nig. Soc. for Anim. Prod. Univ. of Ibadan, Nigeria. 2010;155-158.

58. Adjanohoun E. Some aspects of the sexual cycle of aulocode (*Thryonomys swinderianus* Temminck) and practices on conseequences conducted elevages. The first International Conference The Aulocodicuture: Achievements and perspective. Schrage, R. and Yewaden L.T. eds. 1993; 111-118.

59. Onadeko SA, Amubode FO. Reproductive indices and performance of captive reared grasscutter (*Thryonomys swinderianus* Temminck). Nig. J. Anim. Prod. 2002;29(1):142-149.

60. Amubode FO. Basic information for captive rearing of grasscutter. The Conservation. 1991;1:33-34.

61. Massanyi P, Jancova A, Uhrin V. Morphometric study of male reproductive organs in the rodent species *Apodemus sylvaticus* and *Apodemus flavicollis*. Bull. Vet. Inst. Pulway. 2003;47:133-138.

62. Dyce KM, Sack WO, Wensing CJC. Textbook of Veterinary Anatomy. 2nd Ed. W.B. Saunders; 2002.

63. Olukole SG, Oyeyemi MO, Oke BO. Gross anatomy of male reproductive organs of the domesticated grasscutter (*Thryonomys swinderianus* Temminck). Proc.of 25th Ann. Conf. of Nig. Soc. For Anim. Prod. Univ. of Ibadan, Nigeria. 2010;268-271.

64. Olukole SG, Oyeyemi MO, Oke BO. Biometrical observations on the testis and epididymis of the domesticated adult African great cane rat (*Thryonomys swinderianus*). Eur. J. Anat. 2009;13(2):71-75.

65. Adebayo AO, Oke BO, Akinloye AK. The morphology and morphometry of the male accessory sex gland in the greater cane rat (*Thryonomys swinderianus* Temminck). Book of Abstract of the 46th Ann. Congress of the Nig. Vet. Med. Assos. (NVMA). 2009;6.

66. Yehoah S, Adamu EK. The cane rate. Biologist. 1995;42(2):86-87.

67. Adu EK. Dexieme Conference on promoting broadcast from the breeding of the cane rats in Africa. Su DuSAHARA; 2002.

68. Juma N, Ondwasy HO. Improved management of indigenous chicken sustainable technologies contributing to the socio-economic welfare of rural household. Proc. Of the 8th Kenya Agricultural Research Institute Biennial Scientific Conference, Nairobi, Kenya. 2002;359-364.

69. Benjamin UU, Akinyemi AF, Ijeomah HM. Problems and prospects of grasscutter (Thryonomys swinderianus) farming in Ibadan, Nigeria. J. Agric. Forestry and the Social Sci. (JOAFSS). 2006;4(2):24-32.

70. Taiwo AA, JO Fayenuwo, AJ Omole, AK Fajimi, JB Fapohunda, EA Adebowale. Supplementary effect of concentrate feed on the performance of cane rats fed basal diet of elephant grass. Nig. J. Anim. Prod. 2009;36(1):153-160.

71. Saha A, Love HA, Schwart R. Adoption of emerging technologies under output uncertainty, American Journal of Agriculture Economics. 1994;76:836-846.

72. Fakoya EO, Sodiya CI, Alarima CI, Omoare AM. Information needs of farmers in grasscutter production for improving performance in Ona Ara Local Government Area of Oyo State. Proc. of 33rd Ann. Conf. of Nig. Soc. for Anim. Prod. Ayetoro, Ogun State, Nigeria. 2008;300-301.

73. Kilungo JK, Mghenyi E. Factors limiting beef productivity and marketing in Kenya. Working Paper, Tegoneo Institute, Egerton Univ., Nairobi, Kenya; 2001.

74. Ozowa VN. Information needs of small scale farmers in Africa: The Nigerian Example. Quarterly Bulletin of the International Association of Agricultural Information Specialists. IAALD? CABI. 1995;40:1.

75. Gueye EF. (2009). The role of networks in information dissemination to family poultry farmers. World's Poultry Science Journal. 2009;65:115-124.

76. Hill D. (1992). Cattle and Buffalo Meat Production in the Tropics. Longman Scientific and Technical. 1992;210.

77. Majiyagbe KA, Lamorde AG. Nationally cordinated research programme on livestock disease: Subsectoral goals, performance and medium-term research plan. Tropical Veterinary. 1997;15:75-83.

78. Opara MN, Fagbemi BO. Therapeutic effect of Berenil in experimental murine trypanosomiasis using stocks isolated from apparently healthy wild grass-cutters. (*Thryonomys swinderianus*). Proc. Intl. Conf. on Global Food Crisis, April 19th-24th, Owerri, Nigeria. 2009;31-37.

79. Onyeanusi AE, Famoyin JB. Health care management of grass-cutter in captivity: Assessment of causes of mortalities among rearing stock in Ibadan Metropolis. J. of Forestry Research and Management. 2005;2:58-66.

80. Awah-Ndukum J, Tchoumboue, Tong JC. Stomach impaction in grass-cutter (*Throyonomys swinderianus*) in captivity: Case Report. Trop. Vet. 2001;19(2): 60-62.

81. Opara MN, Fagbemi BO. Haematological and plasma biochemistry of the adult wild African grass-cutter (*Thryonomys swinderianus*). Azoonosis factor in the tropical humid rain forest of southeast Nigeria. Ann. N. Y. Acad. Sci. 2008;1149:394-397.

82. Yeboah S, Simpson PK. A preliminary survey of ecto and endo parasites of the grass-cutter (*Thronomys swinderianus* Temminck) case study in Ekumfi Central Region of Ghana. J. of the Ghana Sci. Assos. 2004;3(3):2-5.

83. BBC News. Scientist find new strain of HIV. 2 August 2009; 2009.

84. Nang-Beifubah AM. 2014.
Available: http://thechronicle.com.gh/bush-meat-not-infested-with-ebola-health-director-assures-nation.

# 3

# Effects of Petroleum Products in Soil on α-Amylase, Starch Phosphorylase and Peroxidase Activities in Cowpea and Maize Seedlings

## F. I. Achuba[1*] and P. N. Okoh[1]

[1]*Department of Biochemistry, Delta State University, PMB 1, Abraka, Nigeria.*

### *Authors' contributions*

*This work was carried out in collaboration between all authors. Author FIA designed the study, wrote the protocol and wrote the first draft of the manuscript. Author PNO reviewed the experimental design and all drafts of the manuscript. Author FIA managed the analyses of the study and performed the statistical analysis. All authors read and approved the final manuscript.*

*Editor(s):*
(1) Moreira Martine Ramon Felipe, Departamento de Enxeñaría Química, Universidade de Santiago de Compostela, Spain.
(2) Mintesinot Jiru, Department of Natural Sciences, Coppin State University, Baltimore, USA.
*Reviewers:*
(1) Anonymous, Senegal.
(2) T. Muthukumar, Root and soil Biology laboratory, Department of Botany, Bharathiar University, India.
(3) Anonymous, Argentina.
(4) Anonymous, Greece.
(5) Olutayo M Adedokun, Crop and Soil Science, University of Port-Harcourt, Nigeria.
(6) Anonymous, Nigeria.

## ABSTRACT

**Aims:** To determine the effect of petroleum products (kerosene, diesel, engine oil and petrol) contaminated soil at various concentrations on the activities of α-amylase, starch phosphorylase in the cotyledons of cowpea and maize seedlings as well as peroxidase activity in the leaves of both seedlings.
**Place and Duration of Study:** This study was conducted in Delta State University, Abraka, Nigeria between April 2007 and August 2011.
**Methodology:** Improved varieties of maize (*Zea mays* L) and *Vigna unguiculata* (L) Walp were planted in soil contaminated at different concentrations comprising six groups. Each group was replicated five times. Groups 1 to 5 contained 0.1%, 0.25%, 0.5%, 1.0% and 2.0% (v/w) respectively of each of the petroleum products while group six served as control (0.0%). Three

seeds were planted in each bag and watered daily. Four days after germination the activities of α-amylase, starch phosphorylase in the cotyledons of the cowpea and maize seedlings were analysed. This was followed by the determination of peroxidase activity in the leaves of cowpea and maize seedlings four, eight and twelve days after germination.

**Results:** The results showed that the petroleum products caused metabolic perturbations in the seedlings. This is indicated by the significant ($P < 0.05$) decrease in the activities of starch degrading enzymes: α-amylase and phosphorylase as well as peroxidase activity compared to their respective control values.

**Conclusion:** Kerosene decreased the activities of the enzymes more than the other petroleum products. The effect of petroleum products contaminated soil was more severe in cowpea seedlings relative to maize seedlings.

*Keywords: α-amylase; starch phosphorylase; peroxidase; cowpea seedlings; maize seedlings; soil.*

## 1. INTRODUCTION

Plant growth and development depend on resources present in soil and air, which consists of external and internal growth factors [1]. Presence of petroleum in the external environment leads to changes in the growth and development pattern of the plant. Petroleum compounds are highly toxic to plants and are detrimental to their growth and development. Petroleum is toxic to higher plants [2-5]. Since seed germination is the first physiological process affected by petroleum, the ability of a seed to germinate in a medium containing petroleum would be indicative of its level of tolerance to this chemical. One toxic effect of petroleum on plant is the depression of seed germination. Various reports hinted that crude oil [6-9], water soluble fraction of crude oil [10] and spent engine oil [11] inhibited seed germination. The retardation in seed germination due to exposure to petroleum had been attributed to decrease in available air and water [12,13]. Vwioko and Fashemi [14] indicated that soaking the seed in water before sowing reduced drastically the effect of spent engine oil on seed germination. Moreover, the inhibitory effect of petroleum on germination was equally attributed to hydrocarbon mediated decrease in nutrient mobilizing enzymes in germinating bean seed [4].

Eriyamremu et al. [6] reported that Bonny light crude oil alters protease activity in cowpea seedlings. In addition, Achuba [4] reported that exposure to whole crude oil inhibited the activities of α -amylase and phosphorylase in the cotyledon of germinating cowpea seeds. Petroleum stress can induce three possible types of metabolic modification in plants. These include alteration in the production of pigments such as chlorophyll [4,5,15], increased production of metabolites such as glucose, total carbohydrate as well as proteins and amino acids [4] and

alterations in plant enzyme activities [4,6] An increase in lipid peroxidation product was observed in plant exposed to petroleum [16,17]. The activity of peroxidase was lower in plant exposed to pollution [18].

The southern region of Nigeria is rich in petroleum resources. The exploitation, processing, transportation as well as disposal of waste oil have resulted in the contamination of the environment [19]. This exposes the biota to the deleterious effects of petroleum pollution. The major occupation of the inhabitants of this region is fishing and farming and some of the main cultivated crops are cowpea and maize [6]. Previous studies have reported that petroleum products penetrate the pore spaces of soil thereby affecting terrestrial vegetation and subsequently impede photosynthesis and other plant physiological processes [16,17,20]. Environmental consequences of refined petroleum products such as kerosene, diesel, engine oil and petrol have not been given the proper recognition they deserve. The aim of the current investigation was to monitor the effects of refined petroleum products on amylase, starch phophorylase and peroxidase activities in cowpea and maize seedlings.

## 2. MATERIALS AND METHODS

### 2.1 Refined Petroleum Products and Planting Materials

The refined petroleum products of known specific gravities (kerosene = 0.81; diesel = 0.85; engine oil=0.87; petrol = 0.75) were obtained from Warri Refining and Petrochemical Company, Warri, Nigeria. Improved varieties of maize *(Zea mays L)* were obtained as single batch from Delta Agricultural Development Project (DTADP) Ibusa Delta State, Nigeria. Improved varieties of *Vigna unguiculata* (L) Walp were obtained from International Institute of Tropical Agriculture IITA,

Ibadan, Nigeria. The soil (sand 84%, silt 5.0%, clay 0.4% and organic matter 0.6%, pH 6.1) was obtained from a fallow land in Delta State University, Abraka. The nutrient content of the soil used is shown in Table 1. The experiment was carried out under laboratory conditions (temperature 28°C and 12hr day/ night).

**Table 1. Physicochemical properties of test soil**

| Parameters | Value |
| --- | --- |
| pH | 6.09 |
| Total organic carbon, % | 2.90 |
| Phosphorus, mg/kg | <0.01 |
| Nitrogen, mg/kg | 8.47 |
| Nitrate, mg/kg | 9.86 |
| Cation exchange capacity, meq /100g | 0.74 |
| Sodium, mg/kg | 9.06 |
| Potassium, mg/kg | 6.72 |
| Calcium, mg/kg | 2.98 |
| Magnesium, mg/kg | 0.31 |

## 2.2 Soil Treatment and Planting of Seeds

One thousand six hundred grams of soil was added to each small size planting bags (1178.3 cm$^3$, 15 cm deep) and divided into six groups of five replicates. Groups 1 to 5 contained 0.1%, 0.25%, 0.5%, 1.0% and 2.0% (v/w) respectively of each of the petroleum products while group six served as control (0.0%).To the first bag, 1.6 ml of kerosene, corresponding to 0.1%, was added. The petroleum product treated soil sample was mixed vigorously with hand to obtain homogeneity of the mixture. The procedure was repeated for all the concentrations and the petroleum products. Each treatment including control was replicated five times. The treatments were watered every day in order to keep the soil moist. The design of the experiment was completely randomized design (CRD) [4].

Damaged seeds were determined by floatation. All seeds that floated on water were discarded and others that remained at the bottom of water were deemed potentially plantable. Three seeds were sown in each test bag to an approximate depth of 2 cm immediately after pollution and kept under partial shade. During the experiment 80 cm$^3$ of water was supplied to the set up as at when needed to keep the soil moist. Germination [which is indicated by the appearance of epicotyls (for cowpea) and hypocotyls (for maize) above the soil level] records was taken at 4 days interval up to 12 days. Seeds, which failed to sprout after 12 days were regarded as not germinable. At the end of each experimental period, the seedlings were carefully removed

from the bags by destroying the bags while the bulk soil containing the seedling was placed under slow running tap water to wash off the soil particles.

## 2.3 Preparation of Extract and Determination of α-Amylase Activity

A 10 ml of 0.05M phosphate buffer, pH 8.0 (prepared in the laboratory) was added to the cotyledon, stored in ice, (0.5g) and homogenized manually by grinding in a ceramic mortar and pestle with 0.5g of acid wash sand. The homogenate was then centrifuged at 3000 g for 15minutes. The supernatant containing the crude enzyme was used to determine the level of activity of α-amylase.

α-amylase assay was carried out by the method of Gupta et al. [21] and the activity calculated by using a formula proposed by Xiao et al. [22]. Assay reaction was initiated by adding 0.5 ml of starch solution (prepared in the laboratory) and 0.5 ml of enzyme in 0.1M phosphate buffer at pH 8.0 and incubated at 37°C for 15 minutes. The reaction was then terminated by adding 1ml of 1 $NH_4Cl$. Then the mixture was then diluted to nearly 9ml with water followed by the addition of 1ml of iodine reagent. Finally the volume was adjusted to 10 ml with distilled water and the intensity of colour development was determined by measuring the absorbance at 620 nm with SP 1800 UV/SP spectrophotometer.

## 2.4 Preparation of Cotyledonary Extract and Determination of Starch Phosphorylase Activity

To the cotyledon sample (1.0g), 10ml of ice cold water was added and homogenized until the formation of thick slurry. The homogenate was filtered through a cheese cloth, then through a Watman filter paper followed by filtration using Buchner funnel and suction. The filtrate was placed in a water bath at 50°C for 5 minutes to inactivate amylase and other enzymes. It was then cooled, followed by the addition of 2 g of cold $NH_4SO_4$ and centrifuged 15000 g to remove precipitated proteins. The supernatant was then decanted and used as the enzyme source.

## 2.5 Determination of Phophorylase Activity

A 5.0 ml of starch solution was placed in each of the 5 cuvettes along with 1 ml of potassium phosphate buffer, pH 5.5. Then 2 ml of the enzyme extract was added to each tube as quickly as possible and allowed to stand for 60 seconds. This was followed by the addition of 1.0M potassium iodide solution (1MKI) and the absorbance of each tube read at 660 nm. The amount of starch in each tube was calculated based on the standard curve obtained earlier.

## 2.6 Determination of Preoxidase Activity

The assay was carried out by the method reported by Rani et al, [23]. The reaction mixture consisted of 3ml of buffered pyrogallol (0.05M pyrogallol in 0.1M phosphate buffer (pH 7.0) and 0.5ml of 1% $H_2O_2$. To this was added 0.1ml enzyme extract and O.D change was measured at 430 nm for every 30 seconds for 2 minutes. The peroxidase activity was calculated using an extinction coefficient of oxidized pyrogallol (4.5 liters/Mol).

## 2.7 Statistical Analysis

The results were expressed as mean $\pm$ SEM. All results were compared with respect to the control. Comparisons between the test and control were made by using Analysis of Variance (ANOVA), Least Significant Difference (LSD) was used to conduct Post Hoc test for the significant difference. Differences at $p < 0.05$ were considered as significant.

## 3. RESULTS

Cotyledons of both cowpea and maize seedlings grown in soil treated with kerosene, diesel, engine oil and petrol showed a reduction in $\alpha$-amylase activity when tested four days after germination (Fig. 1). The reduction in $\alpha$-amylase activity generally increased with increasing concentrations of petroleum products in soil and was greater for cowpea than in maize seedlings. It is evident that kerosene treatment of soil affected both seedlings more than the other petroleum products (Fig. 1).

Cotyledons of both cowpea and maize seedlings grown in soil treated with kerosene, diesel, engine oil and petrol showed a reduction in starch phosphorylase activity when tested four days after germination (Fig. 2). Depression of starch phosphorylase activity generally increased with concentration and was lesser in cowpea than in maize seedlings relative to the other petroleum products.

The activities of peroxidase in the leaves of cowpea and maize seedlings grown in kerosene,

diesel, engine oil and petrol treated soils after four, eight and twelve days of germination are shown in Fig. 3. Generally, peroxidase activity significantly ($p < 0.05$) decreased relative to the control, Moreover, kerosene was more toxic than the other petroleum products and affected cowpea seedlings more than maize.

## 4. DISCUSSION

Petroleum mediated alterations in the activity of plant enzymes were earlier reported [4,24]. In the present study, refined petroleum products inhibited starch degrading enzymes in the cotyledons of germinating cowpea and maize seeds. The dependence of the two starch degrading enzymes in cowpea and maize seedlings on concentration of petroleum products in soil is shown in Figs 1 and 2 respectively. The enzymes: α-amylase and starch phosphorylase are necessary for degrading polysaccharide in seeds. Starch phosphorylase act repeatively on the non reducing end of amylo pectin branches to give glucose-1-phosphate which, after conversion to glucose-6-phosphate, could enter the tricaboxylic acid cycle via glycolytic pathway for production of energy needed by germinating plants [4,25,26]. Therefore, inhibition of starch phosphorylase could disturb respiratory activities of germinating cowpea and maize seedlings as earlier reported [10,10,6,27]. Similarly, α-amylase in conjunction with β-amylase cleaves starch at α- (1→ 4) glycosidic bond from the non reducing end of amylopectin to produce maltose units that are degraded by α-glucosidase to form glucose [28] needed for cellular metabolism [26]. Generally, inhibition of the activities of starch phosphorylase and α-amylase by the refined petroleum products could disturb the production of glucose-1-phosphate and free glucose, thereby predisposing seedling to wide array of metabolic perturbations in cellular metabolism. Comparatively, kerosene seems to inhibit starch phosphorylase and α-amylase activities more than the other three refined petroleum products. The severe toxic effects of kerosene have been attributed to its effects on soil microorganisms [29]. The effect of these refined petroleum products was more pronounced in cowpea compared to maize seedlings. This is in agreement with the report of Coskun and Zihnioglu [30] who showed that monocotyledonous seeds are less affected by toxicants than dicotyledonous seeds.

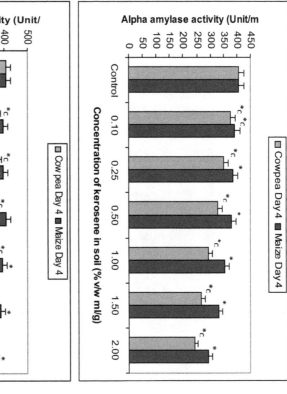
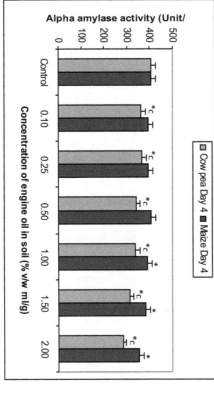
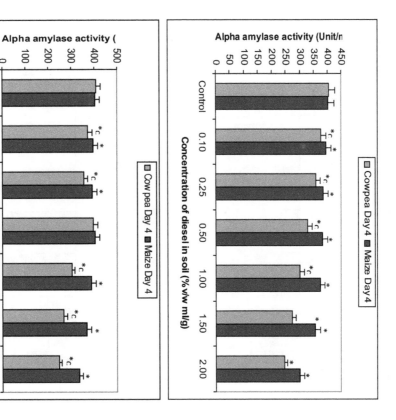

**Fig. 1.** Effect of concentration of petroleum products on α-amylase activities in cotyledom of cowpea and maize after four days of germination. *Significantly lower as compared to control; +Significantly lower as compared to engin oil; ++Significantly lower as compared to kerosene; e Significantly higher relative to control; [a]Significantly lower relative to other petroleum products; [b]Significantly higher relative to other petroleum products; [c]Significantly lower in cowpea relative to maize seedlings; [d]Significantly higher in cowpea relative to maize seedlings

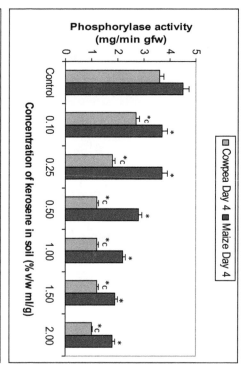

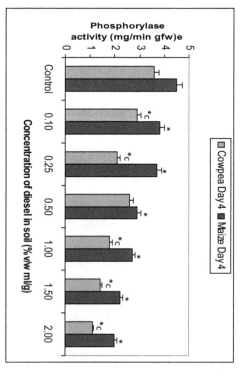

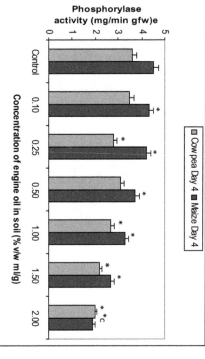

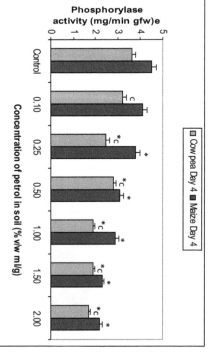

**Fig. 2.  Effect of concentration of petroleum products on phosphorylase activities in cotyledom of cowpea and maize after four days of germination.  \*Significantly lower as compared to control; \*+Significantly lower as compared to engin oil; ++Significantly lower as compared to kerosene; eSignificantly higher relative to control; aSignificantly lower relative to other petroleum products; bSignificantly higher relative to other petroleum products; cSignificantly lower in cowpea relative to maize seedlings; dSignificantly higher in cowpea relative to maize seedlings**

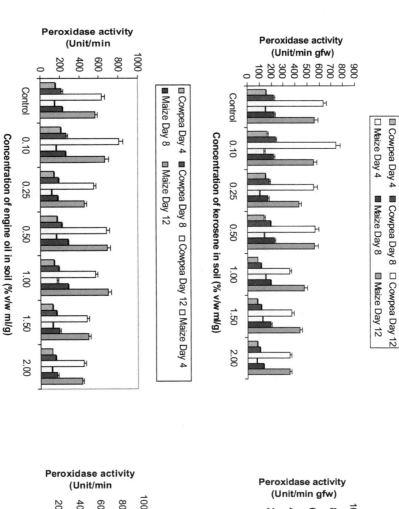

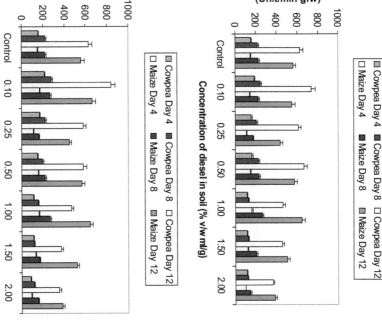

Fig. 3. Effect of concentration of petroleum products on peroxidase activities in leaves of cowpea and maize after four, eight and twele days of germination. No significant difference in peroxidase activity between cowpea and maize (P= 0.87); time had a significant effect on peroxidase activity (P= 0.00); significance difference in activity between control and other concentrations (P= 0.00); significance difference in peroxidase activity among the petroleum products (P= 0.00); time and plant species had significant interaction with peroxidase activity (P= 0.00); time, concentration and petroleum product had significant effect on peroxisae activity (P= 0.00); concentration and petroleum product had significant effect on peroxidase activity (P= 0.00); concentration, plant species, petroleum product had significant interaction (P= 0.03)

Peroxidase activity was lower in the leaves of cowpea and maize seedlings grown in refined petroleum products treated soil after four, eight and twelve days of germination (Fig. 3). This is in contrast to the increase in the enzyme activity in seedlings grown in cotton wool tainted with heavy metals [31]. The decrease in peroxidase activity observed in this study, which was more pronounced at higher concentrations (Fig. 3) of petroleum products in soil, could indicate the initiation of disruption in the biochemical process that precedes the appearance of apparent symptoms of toxicity. That the cowpea and maize seedlings are under metabolic perturbations is further highlighted by the inhibition of peroxidase activity after 12 days of exposure to petroleum product treated soil (Fig 3). Like starch phosphorylase and α-amylase, peroxidase activity was affected more by kerosene compared to other three refined petroleum products. The effect is more pronounced in cowpea compared to maize seedlings (Fig. 3). Peroxidases are a family of enzymes, which are involved in a variety of cellular function such as lignifications, suberization, cell wall elongation, growth, regulation of cell wall biosynthesis and plasticity [32-35]. This explains why seedling exposed to petroleum hydrocarbon exhibited a retarded growth as was observed by earlier reports [4,5]. It has been speculated that peroxidase restricts elongation growth by the formation of diphenyl cross-links. This study, therefore, in part, seems to suggest that one of the mechanisms of petroleum mediated growth retardation is via inhibition of peroxidase activity.

## 5. CONCLUSION

It is pertinent to state that petroleum products mediated toxicity in exposed plants is achieved via inhibition of starch degrading enzymes in the cotyledon of germinating seeds as well as inhibition of peroxidase activity in the leave of the seedling. Moreover, the toxicity of kerosene is more severe than the other petroleum products studied.

## COMPETING INTERESTS

Authors have declared that no competing interests exist.

## REFERENCES

1.   Shanker AK, Carlos Cervantes T, Loza-Tavera H, Avudainayagam S. Chromium toxicity in plants Environ. Int. 2005;31:739-735.

2.   Sparrow SD, Sparrow EB. Microbial biomass and activity in a subarctic soil ten years after crude oil spills. J. Environ. Qual. 1988;17:304-309.

3.   Amadi A, Abbey SD, Nma A. Chronic effect of oil spill on soil properties and michroflora of rainforest ecosystem in Nigeria. Water, Air Soil Pollut. 1996;86:1-11.

4.   Achuba FI. The effects of sublethal concentrations of crude oil on the growth and metabolism of cowpea (Vigna unguiculata) seedlings. The Environmentalist. 2006;26:17-20.

5.   Peretiemo-Clarke BO, Achuba FI. Phyto-chemical effect of petroleum on peanut (Arachis hypogea) seedlings. Plant Pathol. J. 2007;6:179-182.

6.   Eriyamremu GE, Iyasele JU, Osubor CC, Anoliefo GO, Osagie VE, Osa MO. Bonny light crude oil alters protease and respiratory activities of germinating beans (Vigna unguiculata) (L) seeds, J. Sci Eng. Technol. 1999;6(1):1589–1600.

7.   Akaninwor JO, Ayeleso AO, Monago CC. Effect of different concentrations of crude oil (Bonny light) on major food reserves in guinea corn during germination and growth. Sci. Res. Essay. 2007;2(4):127-131

8.   Adedokun OM, Ataga AE. Effects of crude oil and oil products on growth of some edible mushrooms. J. Appl Sci. Environ. Mgt. 2006;10(2):91-93.

9.   Baek K, Kom H, Oh H, Young B, Kim J, Lea I. Effect of crude oil components and bioremediation on plant growth. J. Environ. Sci Health. 2004;A39(9):2465-2477.

10.  Anoliefo GO, Okoloko GE. Comparative toxicity of forcados blends crude oil and its water soluble fraction on seedlings of Cucumeropsis manni naudin. Nig. J.Appl. Sci. 2000;18:39-49.

11.  Sharifi M, Sadeghi Y, Akbarpour M. Germination and growth of six plant species on contaminated soil with spent engine oil. Int. J. Environ. Sci. Tech. 2007;4(4):463-470.

12.  Atuanya EI. Effects of waste engine oil pollution on physical and chemical properties of the soil. Nig. J. Appl. Sci. 1987;55:155-176.

13.  Isirimah NO, Zoufa K, Loganathan P. Effect of crude oil on maize perfomanc and soil chemical properties in the humid forest

zone of Nigeria. Discov. Innov. 1989;1:95-98

14. Vwioko DE, Fashemi DS. Growth response of *Ricinus communis* L. (Castor oil) in spent lubricating oil polluted soil. J. Appl. Environ. Mgt. 2005;9(2):73-79.

15. Malallah GA, Fzal M, Gulsham S, Abraham D, Kunan M, Dhami MSI. Vicia faba as a bio indicator of oil pollution. Environ. Pollut. 1996;92(2):213-217.

16. Achuba FI. Spent engine oil mediated oxidative stress in cowpea (*Vigna unguiculata*) seedlings. EJE FAChe. 2010;9(5):910-917.

17. Nwaogu LA, Onyeze GCO. Effects of spent engine oil on oxidative stress parameters of *Teferia occidentalis* leaves. Nig. J. Biochem. Mol. Biol. 2010;25(2):98-104.

18. Aki C, Guneysu E, Acar O. Effect of industrial waste water on total protein and the peroxidase activity in plants. Afr. J Biotechnol. 2009;8(20):5445-5448.

19. Egborge ABM. Water pollution in Nigeria: Bio-diversity and Chemistry of Warri River. Ben Miller publication, Warri; 1994.

20. Odjegda VJ, Sadiq AO. Effects of spent engine oil on the growth parameters, chlorophyll and protein level of *Amaranthus hybridus* L. The Environmentalist. 2002;22:23-28.

21. Gupta RP, Gigras H, Mohapatra GV, Kumar A, Chauhan B. Microbial amylases: A biotechnological perspective. Proc. Biochem. 2003;38:1599-1616.

22. Xiao Z, Storms R, Tsang A. A quantitative starch-iodine method for measuring alpha-amylase and glutathione activities. Analyt-Biochem. 2006;351:148.

23. Rani P, Meena Unni K, Karthikeyan J. Evaluation of antioxidant properties of berries. India J. Clin. Biochem. 2004;19(2):103-110.

24. Anigboro AA, Tonukari JN. Effect of crude oil on invertase and amylase activities in cassava Leaf extract and germinating cowpea seedlings. Asian J. Biol. Sci. 2008;1(1):56-60.

25. Osagie AU. The yam tuber in storage. Post harvest Research Unit, Department of Biochemistry, University of Benin City, Nigeria.1992;116-117.

26. Chugh LK, Sawhney SK. Effect of cadmium on germination, amylases and rate of respiration on germinating pea seeds. Environ. Pollut. 1996;91(1):1-5.

27. Osubor CC, Anoliefo GO. Inhibitory effect of spent lubrication oil on the growth and respiratory function of *Arachis hypogea* L. Benin Sci. Dig. 2003;1:73-79.

28. McKee T, McKee R. Biochemistry, 2nd Edition. McGraw Hill New York. 1999;345-349.

29. Wemedo SA, Obire O, Ijogubo OA. Myco-Flora of Kerosene polluted soil. Nig. J. Appl. Sci. and Environ. Mgt. 2002;6(1):14-17.

30. Coskun G, Zihnioglu F. Effect of some biocides on glutathione-s-transferase in barley, wheat, lentil and chickpea plants. Turk J. Biol. 2002;26:89-94.

31. Parmar NG, Chanda SV. Effect of Mercury and chromium on peroxidase and IAA oxidase enzyme in the seedlings of *Phaseolus vulgaris*. Turk J. Biol. 2005;29:15-21.

32. Padiglia A, Cruciani E, Pazzaclia G, Medda R, Floris G. Purification and characterization of Opuntia peroxidase. Phytochemistry. 1995;38(2):295-297.

33. Donford HB. Horseradish. Peroxidases: Structure and kinetic properties. Peroxidases in Chemistry and Biology (Everse J, Everse KE, Grisham MB, eds). CRC Press, Boca Raton, Florida USA. 1991;1- 2434. Gaspar T. In: Molecular and Physiological Aspects of plant peroxidase (Greppin H; Penel C; and Gaspar T, eds.) University of Geneva, Geneva. 1986;455.

34. Chanda SV, Singh VD. Changes in Peroxidase and IAA Oxidase Activities During Wheat Grain Development; Plant Physiol. Biochem. 1997;35:245–250.

# 4

# Respiration and Antioxidant Enzymes Activity in Watermelon Seeds and Seedlings Subjected to Salt and Temperature Stresses

**Bárbara França Dantas[1*], Rita de Cássia Barbosa da Silva[2],**
**Renata Conduru Ribeiro[1] and Carlos Alberto Aragão[3]**

[1]*Brazilian Corporation of Agricultural Research, Embrapa Semi-Arid, Petrolina, Pernambuco State, Brazil.*
[2]*Federal Rural University of Pernambuco State- UFRPE, Serra Talhada Campus, Serra Talhada, Pernambuco State, Brazil.*
[3]*Department of Technology and Social Sciences- DTCS, Bahia State University, Juazeiro, Bahia State, Brazil.*

*Authors' contributions:*

*This work was carried out in collaboration between all authors. Authors BFD and CAA designed the study, advised the work, reviewed the experimental design, statistical analysis and reviewed and contributed in all drafts of the manuscript. Authors RCBS and RCR wrote the protocol, managed the analyses of the study and wrote the first draft of the manuscript. All authors read and approved the final manuscript.*

Editor(s):
(1) Masayuki Fujita, Department of Plant Sciences, Kagawa University, Japan.
Reviewers:
(1) Anonymous, Turkey.
(2) Anonymous, Turkey.

## ABSTRACT

This research aimed to evaluate the effect of salt and temperature stress on water uptake and respiration of watermelon seeds during germination process and to quantify changes in the activity of the antioxidant enzymes ascorbate peroxidase (APX), catalase (CAT) and glutathione-S transferase (GST) involved in protection against reactive oxygen species. The research was performed at the Seed Analysis Laboratory (LASESA) of Embrapa Semi-Arid, Petrolina, Pernambuco State, Brazil, from september to december 2011. The experimental design was completely randomized in a factorial 2x3 (cultivars x stress conditions) for respiration evaluation, 3x4

*Corresponding author: E-mail: Barbara.dantas@embrapa.br;*

(cultivars x electrical conductivities) for salt stress assays and 3x3 (cultivars x temperature) for temperature stress assays. The data were submitted to the mean test and evaluated using the standard errors of means. Respiration was measured by $CO_2$ releases by watermelon seeds cv. cv. Crimson Sweet and Charleston Gray evaluated by an infrared gas analyzer, from 0-120 hours of seed imbibition in different environmental conditions (0 $dSm^{-1}$/25°C, 0 $dSm^{-1}$/30°C, 4 $dSm^{-1}$/25°C). The antioxidant enzymes ascorbate peroxidase (APX), catalase (CAT) and glutathione-S transferase (GST) were evaluated in cvs. Crimson Sweet, Charleston Gray and Fairfax seeds and seedlings after five days imbibition in different electrical conductivities (0, 4 and 6 $dSm^{-1}$) or temperatures (20, 25, 30°C). Crimson Sweet seed respiration rate was increased with increasing temperature, salinity however did not influence the respiration of seeds until the radicle protrusion. The activities of APX and CAT enzymes were antagonistically influenced stresses. The activity of GST was not altered with increased electrical conductivity, however high temperatures led to increase of its activity in watermelon seedlings. The antioxidant detoxification system was activated when imposing temperature and salt stress in all studied watermelon cultivars. Different cultivars of watermelon show different adaptation to salt and temperature stress.

Keywords: Climate change; heat; NaCl; cucurbit; metabolism.

## 1. INTRODUCTION

The Fifth Assessment Report of the Intergovernmental Panel on Climate Change [1] indicates that there is a very high confidence that anthropogenic greenhouse gas emissions have caused global warming. This warming causes greater atmospheric dynamics, accelerating the hydrologic and energy cycles in the atmosphere, which consequently can affect the frequency and intensity of extreme climatic events. These climate projections, released by the IPCC, have shown drought scenarios and extreme rainfall events in large areas of the planet. At Brazil, the most vulnerable region to climate change, from a social and agricultural point of view, is the countryside of Brazilian Northeast, where the climate is semi-arid, BSWh' [2] and vegetation is xerophytic corresponding to Caatinga biome. Rainfall reductions appear in most global IPCC models [3], as well as a warming of up to 3-4°C for the second half of the XXI century. This leads to 15-20% flow rate reductions of São Francisco River (main water resource in this region), as well as, dams level reduction and increased salinization of soils and wells [4,5]. This may result in plants temperature, water and salt stresses.

Abiotic stresses such as high temperatures, water deficit and salinity, individually or associated lead to a series of morphological, physiological, biochemical and molecular changes that adversely affect plant growth and productivity [6]. In seeds and seedlings of many species temperature and/or salt stresses may cause a decrease of water uptake by seeds [7],

germination percentage and/or speed [8,9,10,11], changes in reserve mobilization [12], altered antioxidant enzymes activity [13] and different gene expression [14,15]

Drought, salinity, extreme temperatures and oxidative stress are often interrelated and can cause similar cellular damage. Stress in seeds and seedlings causes changes in growth conditions which affects homeostasis in cells metabolism. Thus cells require an adjustment of metabolic pathways, in order to acquire a new state of homeostasis, resulting in acclimation or tolerance [16,17]. Enzymes such as catalase (CAT, E.C. 1.11.1.6) and ascorbate peroxidase (APX, EC 1.11.1.11) and glutathione-S transferase (GST,EC 2.5.1.13) as well as non-enzymatic compounds comprise effective antioxidant systems to protect against oxidative stress [18].

This study aimed to evaluate the imbibition, respiration and changes in the activity of the antioxidant enzymes ascorbate peroxidase, catalase and glutathione-S-transferase involved in protection against reactive oxygen species (ROS) in salt stress and temperature.

## 2. MATERIALS AND METHODS

The research was performed at the Seed Analysis Laboratory (LASESA) of Embrapa Semi-Arid, Petrolina, Pernambuco State, Brazil. Three watermelon cultivars were studied Charleston Gray, Fairfax and Crimson Sweet of 2010/2011 harvest seeds.

## 2.1 Experimental Design and Treatments

Three assays were performed in a completely randomized experimental design and in factorial scheme. For seed respiration evaluation a factorial scheme 2x3 (cultivars x environmental conditions), with two replications of 50 seeds, was arranged with two watermelon cvs. Crimson Sweet and Charleston Gray subjected to three environmental conditions (EC), combining different electrical conductivities and temperatures, which were (1) 0 $dSm^{-1}$ EC at 25°C, (2) 0 $dSm^{-1}$ EC at 30°C, (3) 4 $dSm^{-1}$ EC and 25°C.

Antioxidant enzyme activity was evaluated in three watermelon cvs. Crimson Sweet, Charleston Gray and Fairfax seeds and seedlings after five days imbibition in different electrical conductivities (0, 4 and 6 $dSm^{-1}$)in a 3x4 (cultivars x electrical conductivities) factorial scheme for salt stress assay and in differenttemperatures (20, 25, 30°C) in a 3x3 (cultivars x temperature) factorial scheme for temperature stress assays. All assays were performed in triplicates of 50 seeds.

The data obtained in all three assays were submitted to the mean test and evaluated using the standard errors of means.

## 2.2 Seeds Imbibition and Respiration

Two replications of 50 seeds of watermelon cv. Charleston Gray and Crimson Sweet, were sowed onto rolls of germitest paper soaked in distilled water in a volume equivalent to 2,5 times the paper weight and incubated to germinate in different environmental conditions (0 $dSm^{-1}$/25°C, 0 $dSm^{-1}$/30°C, 4 $dSm^{-1}$/25°C). Quiescent seeds, as well as imbibed for 6, 24, 48, 72, 96 and 120 hours, were weighted evaluation of weight gain through water uptake. Respiration was measured by $CO_2$ releases by watermelon seeds in a 500 $cm^3$ volume chamber linked to a infrared gas analyzer, model IRGA LI-6200 (Li-Cor, Lincoln, Nebraska, USA). An average of 15 measurements, performed at each 5 minutes, was divided by the dry weight of the seeds.

## 2.3 Antioxidant Enzymes Assay

Antioxidant enzymes activity was assayed in seeds of watermelon cv. Crimson Sweet, Charleston Gray and Fairfax. For determination of seeds antioxidant enzymes response to salt stress, triplicates of 50 seeds were sowed onto rolls of germitest paper soaked with different NaCl solutions with electrical conductivity of 0, 4 and 6 $dS.m^{-1}$ [18,19] in a volume equivalent to 2,5 times the substrate paper weight and incubated at 25°C for germination.

For determination of the antioxidant enzymes response to temperature stress, triplicates of 50 seeds were sowed onto rolls of germitest paper soaked in distilled water in a volume equivalent to 2,5 times the substrate paper weight and incubated at 20, 25 and 30°C for germination.

A minimum of 20 seeds and seedlings per replication were collected in liquid nitrogen at the fifth day after sowing for subsequent extraction and quantification of enzyme activity. Catalase (CAT, E.C. 1.11.1.6) and ascorbate peroxidase (APX, EC 1.11.1.11) and glutathione-S transferase (GST,EC 2.5.1.13) were extracted and their activity in germinating watermelon seed was quantified [13]. All assays were performed in triplicates of 50 seeds.

## 3. RESULTS AND DISCUSSION

Watermelon seeds imbibition curve showed a rapid initial water uptake (phase I), that in few hours, reached germination phase II, also called lag phase, due to a plateau in water uptake. Seeds initial radicle protrusion (1%) started at 72 hours (arrow) and an increase in water uptake (phase III) started after 96 hours imbibition (Fig. 1A and 1B). Both cultivars showed a triphasic imbibition curve, as expected, but cv. Charleston Gray seeds showed higher amount of water uptake (Fig. 1B).

Water uptake was very little influenced by mild stresses imposed to watermelon seeds (Fig. 1). Although environmental stresses did affect radicle protrusion in watermelon cultivars Crimson Sweet, Charleston Gray nor Fairfax (data not shown), Crimson Sweet seeds showed a slight increase in water uptake when subjected to supra optimal temperature (Fig. 1A), as well as a high increase in $CO_2$ release, due to respiration (Fig. 1C).

Seeds respiration followed the imbibition curve, however cv. Crimson Sweet seeds respiration was widely affected by temperature increase, mainly during germination phase II (Fig. 1C). Charleston Gray seeds water uptake and respiration was less influenced by temperature and salt stress than Crimson Sweet seeds (Fig. 1).

Seeds of both cultivars subjected to stressful conditions during 0, 6, 24, 48, 72, 96 and 120 hours showed progress of respiratory activity. According to results, it can be estimated that respiration is almost zero when the quiescent seeds have low moisture content, such as 9,54% and 8,75% in Charleston Gray and Crimson Sweet seeds respectively. The seeds respiration enhances quickly reaching high values upon soaking.

Watermelon seeds of the three studied cultivars showed around 100% radicle protrusion in all temperatures to which they were submitted, however seedling development was hindered by 30°C (data not shown) due to a mild temperature stress. A major hydrogen peroxide detoxifying system in plant cells is the ascorbate-glutathione cycle, in which, ascorbate peroxidase (APX) enzymes play a key role catalyzing the conversion of $H_2O_2$ into $H_2O$, using ascorbate as a specific electron donor. The APX responses are directly involved in the protection of plant cells against adverse environmental conditions [20]. Electrical conductivity and temperature affected APX activity in five days old watermelon seedlings. It is noted that APX activity in cotyledons and embryonic axis, had decreased with increasing salinity levels for all cultivars studied (Figs. 2A and 2B). Cotyledons and

embryonic axis of all watermelon cultivars showed the same response to temperature stress, with higher activity in optimum germination conditions and decrease according to stress imposition (Figs. 2C and 2D). These results demonstrate the existence of differential regulation in gene expression [14, 15] and correlated high levels of antioxidant enzymes, such as APX, with the increase of heat stress tolerance in cucumber plants [21].

Environmental stresses cause either enhancement or depletion of CAT activity, depending on the intensity, duration, and type of the stress [22,23,24]. In general, stresses that reduce the rate of protein turnover also reduce CAT activity. Stress analysis revealed increased susceptibility of CAT-deficient plants to paraquat, salt and ozone, but not to chilling [25]. In salt stress there is a maximum activity of catalase (CAT) at cotyledons and embryos at 4.0 dS.m$^{-1}$, in all watermelon cultivars studied. Higher salt concentrations inhibited CAT activity (Figs. 3A and 3B). Increasing temperature enhanced CAT activity in cotyledons and embryos of all watermelon cultivars, although Crimson sweet seedlings showed much higher CAT activity than other two cultivars (Figs. 3C, D).

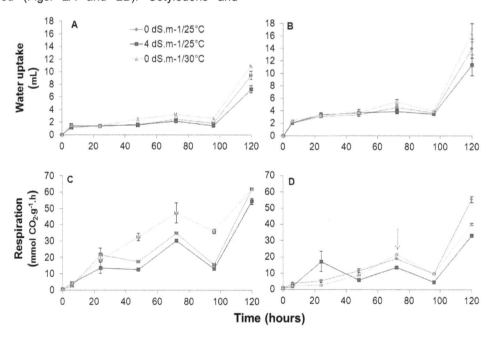

**Fig. 1. Water uptake (A, B) and respiration (C,D) of watermelon seeds cultivars. Crimson Sweet (A, C) and Charleston Gray (B,D) subjected to different conditions of salt and temperature stress**
*Vertical bars represent the standard error of means*

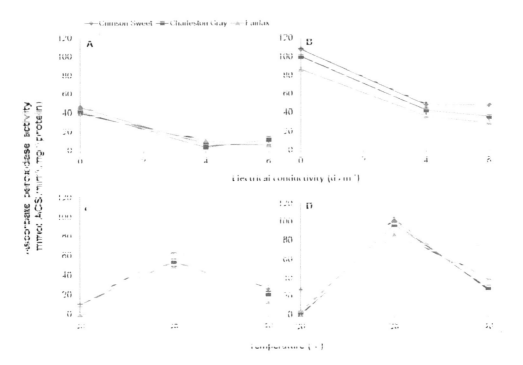

**Fig. 2. Ascorbate peroxidase activity in cotyledons (A,C) and in embryonic axis (B,D) of five days germinated watermelon seeds cultivars Crimson Sweet, Charleston Gray and Fairfax, subjected to different conditions of salt and temperature stress**
*Vertical bars represent the standard error of means*

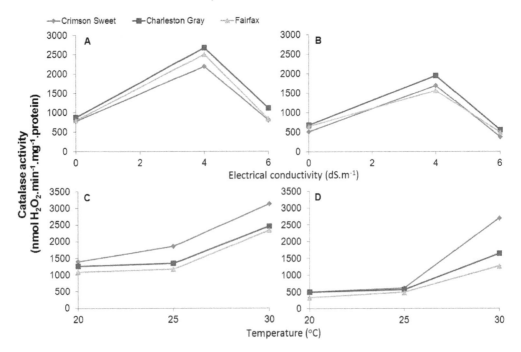

**Fig. 3. Catalase (CAT) activity in cotyledons (A,C) and in embryonic axis (B,D) of five days germinated watermelon seeds cultivars Crimson Sweet, Charleston Gray and Fairfax, subjected to different conditions of salt and temperature stress**
*Vertical bars represent the standard error of means*

GSTs are a group of soluble proteins found in the cytosol, are regulated in vivo by reactive oxygen species (ROS), whose main function is to catalyze the conjugation of reduced glutathione (GSH) with a wide variety of cytotoxic molecules produced as a result of oxidative stress [26]. In this study, temperature influenced GST activity, which showed similar behavior for all studied cultivars and organs and was increased at 30°C (Figs. 4C and 4D), however, none of the cultivars seemed to respond to salt stress by detoxifying mechanism by GST (Figs. 4A and 4B).

It is well known that the respiratory process is the first metabolic activity rapidly activated after seed imbibition, initiating the germination process [27]. Thus, increased release of $CO_2$ characterizes integrity of cellular membranes, including mitochondrial and is indicative of higher seed and seedling ability of reorganization of cell systems and therefore greater vigor. There are many potential sources of ROS in plants, some are reactions involved in normal metabolism, such as photosynthesis and respiration. This makes ROS unavoidable byproducts of aerobic metabolism. Other sources of ROSs belong to pathways enhanced during abiotic stresses, such as drought, high temperature and salinity [16].

During the germination of seeds, several enzymes are involved in metabolic reactions of synthesis and degradation of molecules. Also a part of the germination process, during imbibition occurs the activity of free radical scavengers or antioxidant enzymes which are efficient in the detoxification defense mechanisms [28]. Under normal physiological conditions, a balance between production and elimination of ROS can be disturbed by adverse environmental factors. With increasing stress, ROS formation is enhanced and its elimination should occur steadily to prevent oxidative stress [29]. Thus, the synchronized action of enzymes responsible for the removal of ROS, such as APX, CAT and GST (Figs. 2, 3, 4), confers increased tolerance plants under stress conditions [30]. It has been reported that the production of ROS during seed germination is active and is in fact a beneficial biological activity associated with high germination capacity and development of vigorous seedlings [31]. The germination of seeds appears to be linked to the accumulation of a critical level of $H_2O_2$, suggesting that there is a differential regulation of ROS production and disposal mechanisms in different seed [32].

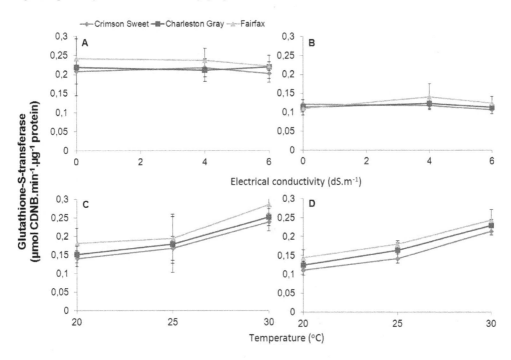

**Fig. 4. Glutathione-S-transferase activity in cotyledons (A,C) and in embryonic axis (B,D) of five days old watermelon seedlings of cultivars Crimson Sweet, Charleston Gray and Fairfax, subjected to different conditions of salt and temperature stress**
*Vertical bars represent the standard error of means*

Although the cultivars responded similarly to temperature and salt stress, Crimson Sweet seeds showed higher CAT activity as well as higher respiration rates than other cultivars (Figs. 1C and 3C). These different results suggest that the positive or negative effects of a particular stress combination could be dependent on the particular plant genotype, species, and/or timing and intensity of the different stresses involved [33].

## 4. CONCLUSION

The antioxidant detoxification system was activated when imposing temperature and salt stress in watermelon cultivars Crimson Sweet, Charleston Gray and Fairfax, allowing adjustment of cell functions during mild stresses. One the other hand, higher stresses, such as 6 $dS.m^{-1}$, may have deleterious effects on these seedlings, especially regarding detoxification by CAT.

The different watermelon did not show same response to the stresses imposed, suggesting different mechanisms of adjustment and tolerance to abiotic stresses.

Further researches must be performed regarding other antioxidant enzymes and reactive oxygen species responses to each cultivar studied in this work, as well as other important watermelon cultivars.

## COMPETING INTERESTS

Authors have declared that no competing interests exist.

## REFERENCES

1. Intergovernmental Panel on Climate Change (IPCC). Climate Change: Impacts, Adaptation, and Vulnerability, fifth assessment report; 2014. Accessed 10 December 2014.
   Available: http://www.ipcc.ch/index.htm
2. Köppen W. Climatology: A study of the climates of the earth. Fondo de Cultura Econômica. México; 1948. Spanish.
3. Angelotti F, Sá IB, Melo RF. Climate change and desertification in the Brazilian semiarid. In: Angelotti F, Sá IB, Menezes EA, Pellegrino GQ. Climate change and desertification in the Brazilian semiarid. Petrolina: Embrapa Semiárido, Campinas: Embrapa Informática Agropecuária; 2009. Portuguese.
4. Cavalcante LF. Salts and its problems in irrigated soils. Centro de Ciências Agrárias/ Universidade Federal da Paraíba. 2000;71. Portuguese.
5. Gondim TMS, Cavalcante LF, Beltrao NEM. Global warming: salinity and consequences on plant behavior. Rev. Bras. Oleag. Fibr. 2010;14(1):37-54. Portuguese.
6. Wang J, Zhang H, Allen RD. Overexpression of an Arabidopsis peroxisomal ascorbate peroxidases gene in tobacco increases protection against oxidative stress. Plant Cell Physiol. 1999;40(6):7225-732.
7. Nizam I. Effects of salinity stress on water uptake, germination and early seedling growth of perennial ryegrass, African J Biotechnol. 2011;10(51):10418-10424.
8. Dantas BF, Ribeiro RC, Matias JR, Araujo GGL. Germinative metabolism of Caatinga forest species in biosaline agriculture. J Seed Sci. 2014;36(2):194-203.
9. Silva FFS, Dantas BF. Effect of temperature on seed germination Sideroxylon obtusifolium (Sapotaceae) of different origins. Rev Sodebras. 2013; 8(90):40-43. Portuguese
10. Oliveira GM, Matias JR, Ribeiro RC, Barbosa LG, Silva JESB, Dantas BF. Seed germination of native tree species of the Caatinga at different temperatures. Scientia Plena. 2014;10(4):1-6. Portuguese.
11. Oliveira GM, Matias JR, Dantas BF. Optimum temperature for germination of native seeds of Caatinga.Informativo ABRATES. 2014;24(3):44-47. Portuguese.
12. Ribeiro RC, Dantas BF., Pelacani CR. Mobilization of reserves and germination of seeds of Erythrina velutina Willd. (Leguminosae - Papilionoideae) under different osmotic potentials. Rev Bras Sementes. 2012;34(4):580-588.
13. Ribeiro RC, Matias JR, Pelacani CR, Dantas BF. Activity of antioxidant enzymes and proline accumulation in Erythrina velutina Willd. seeds subjected to abiotic stresses during germination. J Seed Sci. 2014;36(2):231-239.
14. Xi DM, Liu WS, Yang GD, Wu CA, Zheng CC. Seed-specific overexpression of antioxidant genes in Arabidopsis enhances oxidative stress tolerance during

germination and early seedling growth. Plant Biotech J. 2010;8:796-506.

15. Yamaguchi-Shinozaki K, Shinozaki K. Gene networks involved in drought stress response and tolerance. J Exp Bot. 2007;58(2):221–227.

16. Mittler R. Abiotic stress, the field environment and stress combination. Trends Plant Sci. 2006;11(1):15–19.

17. Suzuki N, Mittler R. Reactive oxygen species and temperature stresses: A delicate balance between signaling and destruction. Physiol Plant. 2006;126(1): 45–51.

18. Perl-Treves R, Perl A. Oxidative stress: An introduction. In: Inzé D, Van Montagu M (eds). Oxidative Stress in Plants. London: Taylor & Francis; 2002.

19. Richards LA. Saline and sodic soils. México: Instituto Nacional de Investigaciones Agrícolas; 1980.

20. Caverzan A, Passaia G, Rosa SB, Ribeiro CW, Lazzarotto F, Margis-Pinheiro M. Plant responses to stresses: Role of ascorbate peroxidase in the antioxidant protection. Genet Mol Biol. 2012;35(4 Suppl):1011-1019.

21. Kang MH, Saltveit EM. Effect of chilling on antioxidant enzymes and DPPH-radical scavenging activity of high and low-vigour cucumber seedling radicles. Plant Cell Environ. 2002;25:1233-1238.

22. Chao H, Qing L, Yan Y. Short-term effects of experimental warming and enhanced ultraviolet-B radiation on photosynthesis and antioxidant defense of *Piceaasperata* seedlings. Plant Growth Regulation. 2009;58(2):153-62.

23. Sharma P, Jha AB, Dubey RS, Pessarakli, M. Reactive Oxygen Species, Oxidative Damage, and Antioxidative Defense Mechanism in Plants under Stressful Conditions. Journal of Botany. 2012:1-26. DOI:10.1155/2012/217037.

24. RagabMoussa H, Abdel-Aziz SM. Comparative response of drought tolerant and drought sensitive maize genotypes to water stress. Aust. J. Crop Sci. 2008;1(1): 31-36.

25. Willekens H, Chamnongpol S, Davey M, Schraudner M, Langebartels C, van Montagu M, et al. Catalase is a sink for $H_2O_2$ and is indispensable for stress defence in C-3 plants. EMBO J. 1997;16(16):4806–16.

26. Bibi N, Yuan S, Zhu Y, Wang X. Improvements of fertility restoration in cytoplasmic male sterile cotton by enhanced expression of glutathione S-transferase (GST) gene. J. Plant Growth Reg. 2014;33(2):420-429.

27. Höffs A, Schuch LOB, Peske ST, Barros ACSA. Effect of seed quality and seeding rate on yield and industrial quality rice. Rev. Bras.Sementes. 2004;26:55-62. Portuguese.

28. McDONALD MB. Seed deterioration: physiology, repair and assessment. Seed Science and Technology. 1999;27(1):177-237.

29. Scandalios JG. Oxidative stress: molecular perception and transduction of signals triggering antioxidant gene defenses. Brazilian Journal of Medical and Biological Research. 2005;38:995-1014.

30. Deuner C, Maia MS, Almeida AS, Meneghello GE. Viability and antioxidant activity in seeds of cowpea genotypes submitted to salt stress. Rev. Bras. Sementes. 2011;33(4):711-720.

31. Liu X, Xing D, LI L, ZHANG L. Rapid determination of seed vigor based on the level of superoxide generation during early imbibition. Photochemical &Photobiological Sciences. 2007;6:767–774.

32. Bailly C. Active oxygen species and antioxidants in seed biology. Seed Science Research. 2004;14(2):93-107.

33. Suzuki N, Rivero RM, Shulaev V, Blumwald E, Mittler R. Abiotic and biotic stress combinations. New Phytol. 2014;203:32–43. DOI: 10.1111/nph.12797.

# 5

# Integrating Mechanical and Chemical Control Treatments to Manage Invasive Weed *Chromolaena odorata* (L.) R. M. King and H. Robinson in Grassland Area

## Muhammad Rusdy[1*]

[1]*Laboratory of Forage Crops and Grassland Management, Faculty of Animal Science, Hasanuddin University, Jl. Perintis Kemerdekaan, Makassar Indonesia, 90245, Indonesia.*

***Author's contribution***

*All this work, from designing the study, writing the protocol, writing the first draft of the manuscript, reviewing the experimental design and all drafts of the manuscript, managing the analyses of the study, identifying the plants, performing the statistical analysis, reading and approving the final manuscript was carried out by author MR.*

<u>Editor(s):</u>
(1) Moreira Martine Ramon Felipe, Departamento de Enxeñaría Química, Universidade de Santiago de Compostela, Spain.
<u>Reviewers:</u>
(1) Asif Tanveer, University of Agri. Faisalabad, Pakostan.
(2) Yasser El-Nahhal, Islamic University Gaza, Palestine.

## ABSTRACT

An experiment was conducted on *Chromolaena odorata* dominated grassland to determine the efficacy of integrated mechanical and chemical control on regrowth of *Chromolaena odorata* and other weeds and to determine their botanical composition at 30, 60 and 90 days after treatment application. Treatments were spraying of glyphosate (Roundup) onslashed *Chromolaena odorata*, spraying of glyphosate on normal *Chomolaena odorata*, spraying of triclopyr (Garlon 4) on slashed *Chromolaena odorata* and spraying of triclopyr on normal *Chromolaena odorata*. Efficacy was assessed on the basis of dry weight of weeds yielded at 30, 60and 90 days after herbicide applications. Both herbicides were more effective when sprayed on normal than on slashed *Chrmolaena odorata*. Regardless of slashing, triclopyr was more effective than glyphosate in suppressing weeds. In glyphosate sprayed plots, *Chromolaena odorata* and other weeds were the dominant plants, whereas in triclopyr sprayed plots, herbage was the dominant plant, however

*Corresponding author: E-mail: muhrusdy79@yahoo.co.id;*

dominance of *Chromolaena odorata* progressively increased over time. The results suggest that the interval between slashing and spraying of herbicides is an important factor to determine the efficacy of integrating slashing and herbicide to control *Chromolaena odorata*.

*Keywords: Chromolaena odorata; integrated slashing and herbicidal control; weed suppression; botanical composition.*

## 1. INTRODUCTION

*Chromolaeana odorata* (L.) King and H. Robinson (hereafter called *Chromolaena*), a species of *Asteraceae* family is a perennial shrub native to subtropical and tropical America [1,2]. It has been reported as one of the world's most invasive species; it is considered to be a serious weed problem in Africa, India, Pacific island and South East Asia [3]. It was introduced to India in the 1840s as an ornamental plant from where it spreads to South East Asia and since the Second World War; it has been spreading rapidly throughout Indonesia. It is considered as the most noxious weed in pasture areas because it reduces grazing area for livestock and hinders biodiversity conservation by changing the botanical composition of pasture [4].

In the pasture area owned by the Faculty of Animal Science Hasanuddin University in Enrekang regency, the weed has covered more than 50% of pasture area thus severely reducing carrying capacity of pasture. Lacks of forage because of reducing carrying capacity generally occur during dry season and during the season; many cattle grazing on the pasture were die because of starvation. *Chromolaena* leaves are not eaten by livestock because it is unpalatable when fed fresh to animals [5]. The weed is also toxic to animals because of high levels of nitrate (5 – 6 times above toxic levels) [6]. As all parts of the plant contain alkaloid that is bitter tasting, livestock will avoid it. Because of these reasons, presence of the weed in grassland area needs to be controlled.

The control of *Chromolaena* is difficult because of its ability to thrive in a wide variety of soils, rapid attainment of reproductive maturity, large production of easily dispersible seeds, a significant proportion of seeds persisting in the soil more than one year and strong ability to resprout after burning [7].

*Chromolaena* can be controlled by mechanical, chemical, cultural, biological and integrated methods. Mechanical control includes uprooting and slashing that have been the most widely

used control measure against the weed. However, to be effective in the long term, the weed must be slashed frequently until its carbohydrates reserve content is exhausted. Escalating labor cost makes this method is prohibitively expensive. Integration with other control methods such as chemical control may be effective and economical.

Many reviews on the use of herbicides for the control of *Chromolaena* are available [8,9]. Herbicides are quicker, cost effective and disturb the soil less where erosion may be of concern. A wide range of herbicides have been evaluated for the control of *Chromolaena*. These include 2,4-D amine, picloram, tebuthiuron, imazapyr, glyphosate and triclopyr. In Indonesia, glyphosate, next to paraquat, are commonly used herbicides to control *Chromolaena* in grassland area. Triclopyr, although effective to control of *Chromolaena* [10], is rarely used in grassland area. There is a paucity of information concerning the efficacy of use of glyphosate and triclopyr in grassland area. The present study was aimed at determining integration of slashing and herbicidal (glyphosate and triclopyr) control method on regrowth suppression of *Chromolaena* and other weeds and observing their botanical composition after treatments applied in grassland area.

## 2. MATERIALS AND METHODS

### 2.1 Study Site

The experiment was conducted during the dry season in a pasture owned by the Faculty of Animal Science Hasanuddin University. The site was located at Maiwa, Enrekang regency South Sulawesi Indonesia from July to November 2012 (3°33'57" S, 119°47'31"E) at about 1300 m above sea level. The climate of the area is tropical monsoon characterized by one rainy season (November to June) and one dry season (July to November). The annual average rainfall was 2426 mm with a daily average temperature of approximately 27.34°C. The soil texture was silty clay loam. The area was heavily infested by combinations of *Chromolaena*, *Stachytarpheta jamaicensis*, *Borreria* sp and some other weeds and herbage species.

## 2.2 Experimental Design and Treatments

The study was conducted on a *Chromolaena* dominated pasture with density of 300 – 500 stems/plot and height 1 – 2 m. Community coefficient values indicated homogeneity among the plant communities were 60 – 70%.The herbicides used were glyphosate and triclopyr. A knapsack sprayer fitted with a fan jet nozzle was used for spraying the herbicides.

The experimental design was a split plot in time design with four integrated chemical and mechanical controls as sub plots and three slashing times after application of herbicide as the main plots. There were three replications for each treatment. The four integrated chemical and mechanical control treatments were: T1 spraying of glyphosate (Roundup) on slashed *Chromolaena* and other plants, T2 spraying of glyphosate on unslashed *Chromolaena* and other plants, T3 spraying of triclopyr on slashed *Chromolaena* and other plants, and T4 spraying of triclopyr on unslashed *Chromolaena* and other plants. Spraying of herbicides was conducted at two weeks after slashing of *Chromolaena* with the slashing height of 10 cm above soil surface. Plot sizes were 5.0 x 5.0 m and a 1 m space between plots was allotted to prevent treatment effects of one plot to other plots. The study area was fenced off using barbed wire to height of 2,0 m to keep out animals and unauthorized persons. The fenced area measured 50 x 40 m. A 100 m wide area outside the fences was ring-weeded using a motorized brush cutter to prevent accidental burning.

Glyphosate and triclopyr were applied at the rates of 2.4 kg a.i./ha and 1.23 kg a.i./ha with concentrations of 10 g and 4 g $L^{-1}$, respectively. The form of triclopyr used was butoxy ethyl ester (Garlon 4). Application of herbicides was conducted on day 15 days after slashing of *Chromolaena*. The efficacy of treatment was determined by measuring dry matter weight of surviving weeds at 30, 60 and 90 days after herbicide application. More dry matter of weeds yielded indicates that the treatment was the less effective. Dry matter weight of regrowth was taken randomly from cutting of plants inside the plots at 10 cm above soil surface in quadrants

measuring 1 m x 1 m. To determine dry matter content, the fresh samples obtained were dried in oven at 70°C for 72 h. The botanical composition was calculated as dry matter yield of species comprising the pasture during experiment.

## 2.3 Statistical Analysis

This experiment was conducted using split plot in time design with three times of sampling (30, 60 and 90 days) as main plot and four integrated mechanical and chemical control treatments as sub-plot. SPSS program version 15 was used to conduct all statistical analysis. Differences among each treatment were analyzed using least significant difference (LSD) method.

## 3. RESULTS AND DISCUSSION

### 3.1 Efficacy of Integrated Chemical and Mechanical Control Treatments

Dry matter yield of weeds at 30, 60 and 90 days after herbicide application (DAHA) are shown in Table 1. Average dry matter yield of weeds sprayed with herbicides on normal *Chromolaena* was significantly lower than when herbicides sprayed on slashed *Chromolaena*. This indicated that the effective treatment to suppress the regrowth of *Chromolaena* and other weeds was spraying of herbicideon normal growth of *Chromolaena* and less effective when the herbicides were applied on slashed *Chromolaena*.

Slashing treatment followed by herbicide application is widely used to control the regrowth of weeds. Slashing of shrub plants reduces their biomass, forces the plants to tap their food reserve in roots or stem base to fuel regrowth and provides more succulent leaves which are more readily penetrated by herbicides. Slashing lowers reserve carbohydrate levels and by timing the herbicide application with the low total non-structural carbohydrate storage, efficacy of herbicide can be maximized [11]. This strategy has been reported to be successfully in suppressing the regrowth of *Chromolaena* in India where 2,4-D herbicideis used [12]. In Swaziland, [13] also reported that slashing followed by spraying of Roundup was more effective in controlling *Chromolaena* than slashing only.

**Table 1. Dry matter yield of *Chromolanea* and other weeds (g/plot) as influenced by integrated mechanical and chemical control methods**

| Treatment | Plants | Days after herbicide application | | | |
|-----------|--------|------|------|------|------|
| | | 30 | 60 | 90 | Mean |
| T1 | *Chromolaena* | 495,00 | 1798,35 | 2053,35 | 1448.90 |
| | Other weeds | 424,00 | 885.85 | 1613.30 | 974.38 |
| | Total | 919.00 | 2684.20 | 3666.65 | 2423.16d |
| T2 | *Chromolaena* | 133.35 | 456.65 | 1067.32 | 552.44 |
| | Other weeds | 310.67 | 647.03 | 1052.70 | 670.13 |
| | Total | 444.02 | 1103.68 | 2120.02 | 1222.57b |
| T3 | *Chromolaena* | 223,35 | 1120.35 | 1430.00 | 924,56 |
| | Other weeds | 663.66 | 1190.00 | 1545.00 | 1132.89 |
| | Total | 887.01 | 2310.35 | 2975.00 | 2057.45c |
| T4 | Chromolaena | 15.00 | 213.35 | 695.65 | 308.00 |
| | Other weeds | 81.35 | 458.26 | 974.35 | 504.65 |
| | Total | 96.35 | 671.61 | 1670.00 | 812.65a |
| Mean | *Chromolaena* | 216.68 | 897.18 | 1311.58 | 808.48 |
| | Other weeds | 369.92 | 795.29 | 1296.34 | 820.52 |
| Mean | Total weeds | 586.88a | 1542.47b | 3257.75c | |

*Mean of total weeds at the same row and column sharing different letter are significantly different (P< 0.05)*

A possible reason why both herbicides were less effective in controlling regrowth of *Chromolaena* in slashed plots was the blocking of downward translocation of absorbed herbicides influenced by the too short interval between slashing of weeds and application of herbicides. This was in agreement with [14] that when the plants are in early flushing, translocation of carbohydrates upward from roots or stem bases to a new flush prevents the downward translocation of foliar applied herbicide to the roots. The maximum height of slashed *Chromolaena* when sprayed with herbicides in this study rarely attained a height of 20 cm and this value might be too low to obtain effective results. [8] Stated that efficacy of various foliar applied herbicides such as triclopyr and glyphosate to *Chromolaena* was high when herbicide was sprayed to actively growing regrowth of 0.5 – 1.0 m tall.

The higher efficacy of both herbicides sprayed on normal *Chromolaena* might be attributed to the high translocation of carbohydrates downward from leaves to roots when herbicides were sprayed, that is after the head and seed had been formed. This was in line with the results of [15] that foliar systemic phloem mobile herbicides have a good efficiency when application were made at post-flowering stages which coincide with translocation of carbohydrates to the roots. This efficiency can reach maximum when the application of herbicides is done in stages where emigration of carbohydrates to the root system is fast.

Average dry matter yields of *Chromolaena* and other weeds in both the slashed and unslashed plots were lower when plants were sprayed with triclopyr than those of glyphosate sprayed plots. The higher efficacy of triclopyr over glyphosate on *Chromolaen* are growth was also reported by [16,17]. [17] reported that by using triclopyr, an acceptable level of control could be obtained with $1.8 - 1.9 \text{ dm}^3/\text{ha}$, whereas by using glyphosate, between 3.5 and 4.3 $\text{dm}^3/\text{ha}$ was required for effective control. This indicated that in grassland area, triclopyr is more suitable to be used to control *Chromolaena* and other weeds than glyphosate. The selective properties of triclopyr give this herbicide is advantage over other herbicides. Vegetation tolerant to triclopyr remains in place and can compete with other plants, increase biodiversity, and reduces the dependency of repeat herbicide application.

## 3.2 Botanical Composition

Botanical composition as influenced by integrated mechanical and chemical control treatments are shown in Fig. 1.

There were 24 species of weeds and herbage recorded in this study. About 80% of total plant species comprised only seven species, namely, *Chromolaena odorata*, *Stachytarpheta jamaicensis*, *Borreria latifolia*, *Borreria laevis*, *Borreria ocymoides*, *Cynodon dactylon* and *Axonopus compressus*.

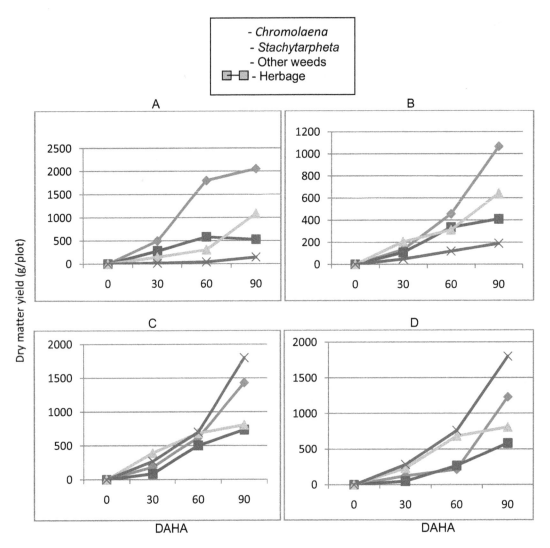

**Fig. 1. Changes in dry matter yield of slashed, glyphosate sprayed plants (A), unslashed, glyphosate sprayed plants (B), slashed, triclopyr sprayed plants (C) and unslashed, triclopyr sprayed plants (D)**

In slashed *Chromolaena* plots sprayed with glyphosate, *Chromolaena* began to be the most dominant species at 30 DAHA and continued to increase until the end of study, but in normal *Chromolaena* plots sprayed with glyphosate, dominance of *Chromolaena* began at 60 DAHA until the end of study. This might be attributed to low efficacy of Roundup sprayed on slashed *Chromolaena* than sprayed on normal *Chromolaena*. In both slashed and unslashed plots, the lowest botanical composition was herbage species. This lowest botanical composition of herbage indicated that glyphosate was unsuitable to control *Chromolaena* in pasture area. This may be attributed to mode of action of glyphosate that is, non selective and

kills all plants, including grasses [18]. Conversely, in triclopyr sprayed plots, herbage was always the most dominant plant, conversely, at 30 and 60 DAHA, botanical compositions of *Chromolaena* were low, comparable to *Stachytarpheta*, however at 60 DAHA, *Chromolaena* was dominant again. The highest botanical composition of herbage in triclopyr sprayed plots may be attributed to differential effects of triclopyr on the regrowth of plant species that reduced regrowth of *Chromolaena* and other herbaceous broad-leaves species but leave grass species unharmed [19]. Thus, spraying triclopyr on *Chromolaena* dominated pasture is very beneficial because it suppresses regrowth of *Chromolaena* and other broadleaf plants but do not kill grasses. However efficacy of Triclopyr was not lasting because at 90 DAHA, *Chromolaena* began to dominate grassland area again this indicated that herbicides application is not lasting and to achieve a 100% success in controlling regrowth, repeated application of herbicide is needed and this makes this method is prohibitively expensive. This was in agreement with [19] that chemical control of *Chromolaena* is not economically feasible and it is unlikely that it would be economic in the extensive grassland.

## 4. CONCLUSION

Spraying triclopyr on normal *Chromolaena* and other weeds is recommended as a suitable control method to suppress the weed in grassland area, because besides providing the highest efficacy of control, it does not kill grasses. However, application of triclopyr and glyphosate on normal *Chromolaena* is not lasting and may require a high cost outlay to achieve a complete control.

## ACKNOWLEDGEMENT

The author is thankful to the Rector of Hasanuddin University for providing financial support and to the Dean of Faculty of Animal Husbandry Hasanuddin University for permitting us to use the Maiwa ranch in Enrekang regency as a siteto carry out this experiment. I would like to thank Muhammad Riadi, a researcher from Faculty of Agriculture Hasanuddin University and the local people for their assistance during the field study.

## COMPETING INTERESTS

Author has declared that no competing interests exist.

## REFERENCES

1. Gautier L. Taxonomy and distribution of tropical weed, *Chromolaena odorata* (L.). King, Robinson H. Candollea. 1992;47:645 – 662.
2. Holm LG, Plucknett DL, Pancho JV, Harberger JP. The World's Worst Weeds, Distribution and Ecology. University Press of Hawaii, Honolulu; 1997.
3. Mc Fadyen RC. *Chromolaena* in South East Asia and the Pacific. Agriculture: New Directions for a New Nation – East Timor (Timor Leste). Ed. by Helder da Costa, Colin Piggin, Cesar J. da James, Fox JJ. ACIAR Proceeding. 2003;113. (Printed Version Published 1n 2003).
4. Mc Fadyen RC. *Chromolaena* in East Timor: History, extent and control. In: *Chromolaena* in Pacific Region. Proc. The 6th International workshop on biological control and management of *Chromolaena odorata* held in Cairns, Australia; 2004.
5. Aro SO, Tewe OO, Aletor VA. Potentials of siam weed (*Chromolaena odorata*) leaf meal as egg colorant for laying hens; 2009. Livestock Research for Rural Development. 2009;21(10).
6. Sajise PE, Palls PK, Norcio NV, Lales JS. Flowering behavior, pattern of growth and nitrate metabolism of *Chromolaena odorata*. Phil. Weed Sci. Bull. 1974;1:17–24.
7. Witkowski ETF, Wilson M. Changes in density, biomass, seed production and seed bank on non-native plant *Chromolaena odorata* along 15 years chonosequence. Plant Ecol. 2001;152(1):13–27.
8. Erasmus DJ. A review of mechanical and chemical control of *Chromolaena odorata* in South Africa. Plant Protection Research Institute, Republic of South Africa; 2014. Accessed on 5 May 2014. Available: www.ehs.cdu.au/chromolaena/proceedings/first/1eras.htm
9. Motooka PL, Ching P, Nagai N. Herbicidal weed control methods for pastures and natural areas of Hawaii. Cooperative Extension Service College of Tropical Agriculture and Human Resources University of Hawaii at Manoa; 2002.
10. Erasmus DJ van Staden J. Screening candidate herbicides in field for chemical control of *Chrmolaena odorata* in South Africa. J Plant Soil. 1987;3:66 – 70.
11. Owens C, Madsen J. Eurasian water milfoil control using contact herbicide phonological timing. Aquatic Plant Control Technical Note CC-01; 1998.
12. Abraham T, Thomas CG, Joseph PA. Herbicides for control of *Chromolaena odorata*. Kerala Agricultural University, India; 2014. Accessed on 23 September 2014. Available:http://www.ehas.cdu.au/chromolaena/proceedings/fourth/abra.htm.
13. Ossom EB, Lupupa S, Mhlongo, Khumalo L. World Agric. Sci. 2007;3(6):704–713.
14. Muniappan R, Marutani M. Mechanical, cultural and chemical control of

*Chromolaena odorata.* Agricultural Experimental Station, University of Guam, USA; 2014.

15. Noureddine RB, Mohammed, Bouchabib B. The evolution on non-structural carbohydrates in the wid jube "Ziziphu*lotus* (L.) Desf." and chemical control strategy. IRACST – Eng. Sci. Technol. 2012;2(6):998–1001.

16. Van Staden J. Chemical control of *Chromolaena odorata*: efficacy of triclopyr and glyphosate applied to regrowth. Appl Plant Sci. 1987;1(1):39–42.

17. Ikuenobe CE, Ayeni AO. Herbicidal control of *Chromolaena odorata* in oil palm. Weed Res. 1998;38(6):397–404.

18. Chang S, Liao CH. Analysis of glyphosate, glufosinate and aminomethylphosphoric acid by capillary electrophoresis with indirect florescence detection. J. Chromatogr. 2002;959:309–315.

19. Ganapathy C. Environmental fate of triclopyr. Environmental Monitoring & Pest Management Branch, Department of Pesticide Regulation, Sacramento, USA; 1997. Accessed 12 January 2014. Available:www.cdpr.ca.gov/docs/emon/pubs/fatememo/ triclopyr.pdf.

# 6

# Screening for Stem Rust (*Puccinia graminis* f. sp Tritici, Eriks. & e. Henn.) Resistance in Mutant Barley (*Hordeum vulgare* l.) Lines

I. J. Obare[1*], M. G. Kinyua[2], O. K. Kiplagat[2] and F. M. Mwatuni[1]

[1]*Kenya Plant Health Inspectorate Service, P.O.Box 49421-00100 Nairobi, Kenya.*
[2]*Department of Biotechnology, University of Eldoret, P.O.Box 1125-30100 Eldoret, Kenya.*

***Authors' contributions***

*This work was carried out in collaboration between all authors. Author IJO designed the study, wrote the protocol and wrote the first draft of the manuscript. Author FMM reviewed the experimental design and all drafts of the manuscript. Authors MGK and OKK managed the analyses of the study. All authors read and approved the final manuscript.*

Editor(s):
(1) T. Muthukumar, Root and Soil Biology Laboratory Department of Botany, Bharathiar University, India.
Reviewers:
(1) Klára Kosová, Division of Plant Genetics, Breeding and Crop Quality, Crop Research Institute, Prague, Czech Republic.
(2) Gyula Oros, PPI HAS, Budapest, Hungary.

## ABSTRACT

Stem rust is a devastating disease in barley that is caused by a fungi (*Puccinia graminis* f. sp tritici, Eriks. and E. Henn). The disease has been controlled for quite some time due to the presence of cultivars carrying the resistant gene *Rpg1*. It has been effective in controlling the various races of stem rust. This was so until the emergence of the race Ug99 from Uganda in the year 1998. This race did break all the resistant genes that were there hence the need to get new sources of resistance. In the current study, mutation breeding was used to create variation for stem rust resistance (Ug99). Seeds of barley (Nguzo variety- $M_O$) were sent to Vienna in Austria for irradiation at the International Atomic Energy Agency at a dosage of 250 gray. The $M_1$ seeds were multiplied in University of Eldoret experimental field. Thousand plants were randomly selected from the $M_1$ population, two ears were harvested of each plant that were subsequently divided within two groups. One group was planted at University of Eldoret experimental field while the other group of a thousand ears were planted at KARI Njoro as $M_2$. Each ear formed a row/line. A susceptible line of wheat was planted as a spreader and inoculated with stem rust -Ug99 in both sites. A total of

183 lines were selected from the two sites. These lines were again replanted in university of Eldoret as $M_3$ in a RCBD design with three replicates in the field to determine adult plant resistance and in the green house in a CRD design to determine the seedling resistance. The non mutated parent, Nguzo was used as a check. The following lines did show resistance both at the seedling level and adult plant level (1, 2, 9, 21, 26, 49, 55, 58, 59, 69, 76 and 78). Mutational breeding is therefore recommended for continual screening of these lines as this race may mutate further.

Keywords: Stem rust (Ug99); mutation.

## 1. INTRODUCTION

Barley is one of the founder crops of Old World agriculture. Genetic markers point to the origin of barley being the Fertile Crescent especially the Israel-Jordan area in the southern part of the Fertile Crescent where there is the highest diversity of the species [1]. This is the area that has the highest probability of being the geographical area within which wild barley was domesticated about 8000 B.C. [2]. The Agriculture Sector, which is the backbone of Kenya's economy and its growth is dependent on increasing barley production amongst other crops and livestock [3]. Barley (Hordeum vulgare L.), is one of the most important cereal crops, in Kenya. Currently about 30,000 ha of land is used for the crop but potential for expansion remains [4]. Rust fungi, responsible for diseases like stem rust, yellow rust and leaf rust is a major contributor to the sub optimal yields realized by farmers. Breeding for resistance has been used as the main method of protection against the fungi. The rust fungi can however overcome host resistance genes and spread new strains through wind dispersal of spores [5]. Stem rust of barley and wheat, (caused by Puccinia graminis f. sp tritici Eriks. & E. Henn.) is historically one of the most important plant diseases. Devastating stem rust epidemics often result in major grain losses [6]. Stem rust "Ug99"-infected plants may suffer up to 100% loss [7] In the current study mutation breeding has been utilised to create variation for resistance. The use of mutagens such as gamma rays for inducing variation is well established. Induced mutations have been used especially by the FAO/IAEA division of nuclear techniques in Agriculture. More than 1800 cultivars obtained either as direct mutants or derived from their crosses have been released worldwide in over 50 countries [8]. The induced mutations aid in development of many agronomic important traits in major crops such as wheat, barley rice and peanuts [9]. The importance of barley as a major crop in Kenya cannot be overlooked because it is the fourth most important cereal in Kenya and the world after maize, wheat and rice [10]. Some of the Major barley producing areas in Kenya includes; Timau, Moiben, Nakuru, on the wetter escarpment of Samburu District near Maralal Town, Molo, and Mau Narok [11]. Barley production in Kenya was estimated to be 75,000MT in 2007 [12]. The aim of this study was to screen for stem rust resistance (Ug99) in mutant barley lines at $M_2$ and $M_3$.

### 1.1 Sites Description

The study was conducted at the University of Eldoret, The geographical coordinates are $0^0$ $30'$ $0''$ North, $35^0$ $15'$ $0''$ East. The site is located 10 Km of Eldoret town, in Uasin Gishu county of enya. It is located at an altitude of 2180 m above sea level; it consists primarily of an agro-ecological zone LH3 [13]. The site is among major wheat growing regions in Kenya. University of Eldoret receives a unimodal rainfall which begins in March. The average annual rainfall range is between 900 mm and 1100 mm and mean annual temperature of 16.6°C. The soils are shallow, ferralsol, well drained, non humic cambisols with low nutrient availability and moisture [13].

The study was also conducted at Kenya Agricultural Research institute, Njoro in Nakuru county, (0º20'S 35º56'E), located in the lower highlands (LH3), at an altitude of 2166 meters above sea level. The temperature ranged between 18-28ºC during the period of study, while the average annual rainfall was about 1,000 mm. The soils are deep, well drained, fertile Vitric Mollic Andosols [13].

### 1.2 Irradiation

Six hundred grams of $M_0$ seeds (non mutated seeds) of the barley variety Nguzo, obtained from East African Maltings in Molo, Kenya were sent to International Atomic Energy Agency in Vienna and subjected to gamma radiation at an iradiation dose of 250 gy (gray) to obtain $M_1$ (mutated seed that gives rise to the first generation of mutants).

## 1.3 Seed Multiplication and Selection

The land to be planted was disc ploughed and harrowed to fine tilth suitable for barley planting. The irradiated $M_1$ seeds were planted in University of Eldoret for seed multiplication. The mutated seed were drilled on a plot measuring 125 m by 40 m. Drills were 5 cm apart. All the agronomic practices like insect pest control, diseases control and weed control were done up to harvest time to ensure good crop establishment. Thousand plants were randomly selected from the $M_1$ population and two ears harvested from each selected plant and divided into two corresponding groups. One group planted at University of Eldoret experimental field while the corresponding group of a thousand ears was planted at KARI Njoro. Each ear formed a line/row.

## 1.4 Planting and Field Management

The $M_2$ seed from each entry were sown in 1M rows. The experimental units were separated by 0.3 m and 0.5 m wide alleyways within and between the blocks, respectively. Sowing was done at an equivalent seeding rate of 125 kg/ Ha. At planting time, Di-ammonium phosphate fertilizer was applied at an equivalent rate of 125 kg/ha. Weeds growth was restricted by applying both pre - and post - emergent herbicides. Stomp® 500 EC (pendimethalin) a broad spectrum, pre-emergent herbicide was applied at an equivalent rate of 2.5 l/ha. At tillering stage (Zadoks' Growth stage 20-29) [13] the plots were sprayed with Buctril MC (bromoxynil + MCPA) at an equivalent rate of 1.5l/ha to control broad-leaved weeds. The trial was top dressed with Calcium Ammonium Nitrate (CAN) at stem elongation stage (Zadoks' GS 30).

## 1.5 Preliminary Data Collection and Selection

Rust development was closely monitored on the test plants and response to rust infection at the adult plant stage was termed "infection response". According to the size of the pustules and associated necrosis or chlorosis infection responses were classified into four categories; R = resistant, MR = moderately resistant, MS = moderately susceptible and S = susceptible [14]. Stem rust severity was assessed using the modified Cobb scale [15]. Entries were evaluated for response to infection and stem rust severity between heading and plant maturity and resistant lines selected and harvested to be advanced to $M_3$. 74 lines were selected from Njoro whereas in University of Eldoret 109 lines were selected. These showed acceptable levels of resistance based on infection type and severity. Each harvested line was harvested and kept in a separate bag to avoid mechanical mixture. The rest of the materials were bulked together.

## 1.6 Experimental Procedure

A field experiment was established on land previously under wheat and was disc ploughed and harrowed to fine tilth suitable for barley planting. Each entry was sown in double rows measuring 0.2 × 0.75 m. The entries were randomly assigned within a block. The experiment was layed out in RCBD replicated three times. The experiment was managed as described above for agronomic management and inoculation done as described for preliminary evaluation.

## 1.7 Seedling Screening of $M_3$ Plants for Resistance to Stem Rust in the Greenhouse

Greenhouse experiment was conducted at the University of Eldoret greenhouse. It was put in a Completely Randomized Design (CRD) with three replicates and repeated three times. The selected 183 lines were evaluated for seedling resistance to Ug99 plus their parent as a check. They were planted in Plastic cups measuring 4 cm diameter by 6 cm height filled with about 200 g of a mixture of soil and sand in a ratio of 3:1. The plastic cups were placed in a greenhouse that was maintained at 23°C and 60% relative humidity (RH). In each pot, 3-4 seeds were planted at approximately 2 cm deep and the cups were then placed in large non-draining trays measuring 60 cm × 240 cm and watered to field capacity. The seedlings were inoculated with urediniospores after 14 days using a hand sprayer when the first leaf was fully developed and placed in dark moist chamber maintained at 100% relative humidity, temperature at 13°C for 18 hours. Thereafter, the seedlings were then transferred to a growth chamber that was maintained at about 22°C day and 20 - 21°C night temperature. After the disease had developed, scoring was done according to Stakman et al. [16].

## 2. RESULTS AND DISCUSSION

The germplasms (mutant barley lines) showed varying levels of resistance to stem rust (Ug99) Table 1. At seedling stage, the infection levels ranged from 0 to 4, whereas at adult plant stage the severity ranged from MR to S. The resistance at seedling stage may probably be because the germplasm had resistance conferred by one single major gene that was broken down at adult plant stage. The genotypes that showed some resistance at adult stage may contain a major single gene that remained resistant at seedling and at adult plant stage or they may have minor genes that are working together to reduce the disease [12]. Mutations are used to study the nature and function of genes which are the building blocks and basis of plant growth and development, thereby producing raw materials for genetic improvement of economic crops [17]. Mutation has been used to produce many cultivars with improved economic value and study of genetics and plant developmental phenomena [18,19]. In the present study gamma rays were used to induce variation in barley lines especially for the gene governing stem rust (Ug99) resistance (*Rpg1*). The induced variations that were brought about by the gamma rays had some molecular basis, i.e., change in the base sequence of the DNA molecule coding for the protein. The change in the base sequence can be through base substitution, base addition or deletions. This could explain the induced variations in terms of Ug99 resistance in the barley lines. There some lines which had low infection type in their seedling screening and were showing resistance to Ug99 in the field at adult plant stage.

**Table 1. Summary of seedling resistance infection type and adult plant resistance of the mutant barley lines at M$_3$**

| Mutant lines | Seedling resistance infection type (IT) | APR (Severity % and infection type) |
|---|---|---|
| 1 | 1 | 5   MR |
| 2 | 2 | 20 MR |
| 5 | 4 | 30 S |
| 7 | 4 | 20 MS |
| 8 | 3 | 20 MS |
| 9 | 2 | 15 MR |
| 21 | 1 | 20 MR |
| 23 | 2 | 20 MS |
| 26 | 2 | 10 MR |
| 27 | 3 | 40 MS |
| 34 | 2 | 35 MS |
| 36 | 2 | 15 MS |
| 41 | 2 | 20 MS |
| 44 | 3 | 10 MS |
| 49 | 2 | 10 MR |
| 54 | 2 | 20 MS |
| 55 | 2 | 25 MR |
| 58 | 2 | 20 MR |
| 59 | 1 | 25 MR |
| 62 | 1 | 5   MR |
| 69 | 1 | 10 MR |
| 76 | 1 | 5   MR |
| 78 | 2 | 20 MR |
| 90 | 3 | 45 MS |
| 95 | 1 | 5   MR |
| 124 | 3 | 20 MS |
| 126 | 2 | 30 MS |
| 130 | 3 | 25 MS |
| 156 | 3 | 20 MS |
| 161 | 3 | 15 MS |
| 163 | 2 | 35   MS |
| 165 | 2 | 20   MR |
| 173 | 2 | 30 MS |
| 184 | 3 | 40   S |

*KEY: S-Susceptible, MS-Moderate Susceptible, MR-Moderate Resistant, R-Resistant; 1.Resistant   2.Resistant to moderately resistant 3.Moderately resistant/moderately susceptible 4 Susceptible*

# 3. CONCLUSION

Mutation by irradiation was successfully applied and generated the much needed variability that inferred resistance to the mutant barley lines at $M_2$ and $M_3$ at the seedling level and adult plant level. The following lines did show resistance both at the seedling level and adult plant level (1, 2, 9, 21, 26, 49, 55, 58, 59, 69, 76 and 78).

# 4. RECOMMENDATIONS

This study recommends the following.

- Continual screening of these lines as this race may mutate further (1, 2, 9, 21, 26, 49, 55, 58, 59, 69, 76 and 78).
- Stabilization of these mutant lines through double haploid techniques and backcrossing to reduce the effects of mutation.
- Agronomic traits should be evaluated on the resistant lines to identify superior lines for release as new varieties for commercial purposes.

# ACKNOWLEDGEMENTS

University of Eldoret-EABL Project for funding.

# COMPETING INTERESTS

Authors have declared that no competing interests exist.

# REFERENCES

1.  Badret A, Muller K, Schafer, Pregi REL, Rabey H, Effgen S, Ibrahim HH, Pozzi C, Rohde W, Salamini F. On the origin and domestication of barley (*Hordeum vulgare*). Molucular Biology and Evolution. 2000;17:499-510

2.  Zohary D, Hopf M. Domestication of plants in the Old World, 2nd edition. Oxford University Press. 1993;278.

3.  GOK. Poverty reduction strategy paper for the period 2001-2004; 2001.

4.  East African Breweries Ltd. The Kenyan Beer Industry. Nairobi, Kenya; 2005.

5.  Hovmøller MS, Justesen AF. Rates of evolution of avirulence phenotypes and DNA markers in North-West European population of *Puccinia striiformis* f. sp. *tritici*. Molecular Ecology. 2007;16:4637-4647.

6.  CIMMYT. Sounding the Alarm on Global Stem Rust. An Assessment of Ug99 in Kenya and Ethiopia and Potential for Impact in neighboring Regions and beyond. 2005;26.

7.  Hildebrant D. Future World Wheat Crops Threatened by Ug99 Stem Rust. Farm and Ranch Guide; 2008.

8.  Ahloowalia BS, Maluszynski M. Induced Mutations-A new paradigm in plant. Euphytica. 2001;118(2):167-173.

9.  FAO. Meeting of the technical subgroup of the expert group on International Economic and Social Classifications. United Nations Department of Economic and Social Affairs Statistics Division. New York, U.S.A; 2004.

10. Kenya Maltings Ltd. Annual Reports. Nairobi, Kenya; 2007.

11. National Cereals and Produce Board. Control and regulation of cereals production, marketing, storage, distribution and standard measures. A reporton barley. Nairobi, Kenya; 2007.

12. Jaetzold R, Schmidt H. Farm management handbook of Kenya. Natural condition and farm management information. Ministry of Agriculture, Kenya in cooperation with Germany Agricultural team (GAT) of the GermanyAgency for technical cooperation (GTZ) Nairobi Kenya. 1983;2.

13. Zadok JC, Chang TT, Konzak CF. A decimal code for the growth stages of cereals. Weed Res. 1974;14:415–421. i0022-0493-93-1-38-b20C.

14. Roelfs AP, Singh RP, Saari EE. Rust diseases of wheat: Concepts and methods of disease management. CIMMYT, Mexico City; 1992.

15. Peterson RF, Campbell AB, Hannah AE. A diagramic scale for estimating rust intensity of leaves and stems cereals. Canadian Journal Research. 1948;26:496-600.

16. Stakman EC, Stewart DM, Loegering WQ. Identification of physiological races of Puccinia graminis var tritici U.S Dept of agriculture research service, U.S.A; 1962.

17. Adamu AK, Aliyu H. Morphological effects of sodium azide on tomato (*Lycopersicon esculentum* Mill). Science World Journal. 2007;2(4):9-12.

18. Van Den-Bulk RW, Loffer HJM, Lindhout WH, Koornneef M. Somaclonal variation in tomato: Effect of explants source and a comparison with chemical mutagenesis. Theoretical Applied Genetics. 1990;80: 817–825.

19. Bertagne-Sagnard B, Fouilloux G, Chupeau Y. Induced albino mutations as a tool for genetic analysis and cell biology in flax (*Linum usitatssimum*). Journal of Experimental Botany. 1996;47:189–194.

# 7

# Comparative Performance of Pyrethrum [*Chrysanthemum cinerariifolium* Treviranus (Vis.)] Extract and Cypermethrin on Some Field Insect Pests of Groundnut (*Arachis hypogaea* L.) in Southeastern Nigeria

Frank Onyemaobi Ojiako[1*], Sunday Ani Dialoke[1],
Gabriel Onyenegecha Ihejirika[1], Christopher Emeka Ahuchaogu[2]
and Chinyere Peace Ohiri[1]

[1]*Department of Crop Science and Technology, Federal University of Technology, P.M.B.1526, Owerri, Imo State, Nigeria.*
[2]*Department of Crop Production and Protection, Federal University Wukari, P.M.B.1020, Wukari, Taraba State, Nigeria.*

## Authors' contributions

*This research was carried out collaboratively by all authors. Author FOO designed the study and wrote the first protocol. All the authors were fully involved in the experiment; field work, laboratory extraction, statistical analysis, managing the literature search and editing the manuscript. All authors read and approved the final manuscript.*

<u>Editor(s):</u>
(1) Marco Aurelio Cristancho, National Center for Coffee Research, Cenicafé, Colombia.
(2) Anita Biesiada, Department of Horticulture, Wroclaw University of Environmental and Life Sciences, Poland.
<u>Reviewers:</u>
(1) Anonymous, Turkey.
(2) John A. Mwangi, Horticultural Research Institute, Kenya Agricultural and Livestock Research Organization, Kenya.
(3) K. Baskar, Bioscience Research Foundation, Chennai, India.
(4) Karamoko Diarra, Department of Animal Biology, Cheikh Anta Diop University (UCAD), Senegal.
(5) Alhassan Idris Gabasawa, Department of Soil Science, Institute for Agricultural Research, Ahmadu Bello University, Nigeria.

*Corresponding author: E-mail: frankojiako@gmail.com;*

## ABSTRACT

Study to evaluate the insecticidal efficacy of pyrethrum, *Chrysanthemum cinerariifolium,* relative to a synthetic insecticide (Cypermethrin 10 E. C) in the control of some field pests of groundnut was carried out at the Teaching and Research Farm, School of Agriculture and Agricultural Technology, Federal University of Technology, Owerri, southeastern Nigeria, from March to November, 2012. The experiment was laid out in a 2 x 4 factorial arrangement fitted into a Randomized Complete Block Design replicated three times. Groundnut seed variety, ICGV-IS 96894 (ICRISAT) was subjected to germination test to ensure viability. Seeds were planted at a spacing of 30 cm x 15 cm (220,000 plants/ha) on 24 (1 $m^2$) beds with 1 m between furrows. Pyrethrum was extracted through a simple replicable procedure and tested at four rates (0.00, 0.25, 0.50, 0.75 g/100 ml of water). Cypermethrin 10 EC was tested at 0.00, 0.50, 1.00 and1.50 ml/100 ml of water. Insecticides application, pest sampling and leaf damage assessments were carried out at 4, 6 and 8 weeks after planting (WAP). Yield measurement parameters (seed weight, shell weights and pod density) were assessed. Major arthropod pests identified were; *Macrotermes bellicosus* Smeathman, *Peridontopyge* spp., *Helotrichia serrata* Fabricius, and *Oedaleus nigeriensis* Uvarov. Cypermethrin and pyrethrum applications reduced pest incidence (3.25, 2.50, 2.10 insects) and (4.09, 3.62, 3.42 insects), respectively, when compared with unsprayed plots (6.35, 6.16, 6.20 insects) at 4, 6 and 8 WAP. Insecticide type had no significant effect on the population of majority of sampled pests. Sprayed plots had less damaged leaves - 2.80 (cypermethrin) and 2.83 (pyrethrum) as against 4.52 leaves in unsprayed plots at 8 WAP. Sprayed plots also had increased fresh pod weight (0.61, 0.52 kg in cypermethrin and pyrethrum sprayed plots, respectively) as against 0.14 kg in unsprayed plots. Seed weights (0.26, 0.22 kg in cypermethrin and pymethrum sprayed plots, respectively) were significantly distinct from the control (0.06 kg).There were no significant differences (P = .05) in the dry pod or shell weights based on insecticide types. The efficacy of the insecticides was dose related as higher rates gave better performances. Pyrethrum compared favourably with cypermethrin in controlling the field insect pests of groundnut and could serve as alternative to synthetic pesticides in the management of these pests in southeastern Nigeria.

Keywords: Arachis hypogaea; cypermethrin; Holotrichia serrata; Macrotermes bellicosus; Oedaleus nigeriensis; Peridontopyge spp; pyrethrum extract.

## 1. INTRODUCTION

Groundnut (*Arachis hypogaea* L.), also called peanut, is a species in the legume or bean family, *Fabaceae.* It is said to have originated from South America, in the region of Bolivia, Argentina and Brazil and is one of the most popular commercial plant groups in West Africa, especially in Northern Nigeria (latitude 10°N). China is the largest producer of groundnut in the world followed by India, USA, Nigeria and Indonesia. In sub-Saharan Africa, groundnut is the 5th most widely grown crop, closely trailing behind maize, sorghum, millet and cassava. Nigeria is the largest producer in Africa, producing 30% of the Continent's total, followed by Senegal and Sudan (each with about 8%), and Ghana and Chad with about 5% each [1].

Groundnut, the third major oilseed of the world, next to soybean and cotton [2], provides a vital source of cash income and nutritious high protein and fatty food. The kernels are eaten raw, roasted, sweetened or processed into peanut butter which is rich in protein and vitamins A and B. They are also consumed as confectionary product. Groundnut oil is edible and is also used in the production of soap, cosmetics, lubricants, olein, stearin and their salts [3]. Groundnuts could prevent child malnutrition and are useful in the treatment of hemophilia, stomatitis and for the prevention of diarrhea. The meal is beneficial for growing children, pregnant women, nursing mothers [4], and is a good source of niacin, which contributes to brain health and blood flow [5].

Some insect pests such as white grubs (*Phyllophaga* spp), termites (*Macrotermes bellicosus* Smeathman), gram pod borers (*Helicoverpa armigera* Hubner), groundnut leaf miners (*Aproaerema modicella* Deventer), Aphids (*Aphis craccivora* Koch.), etc attack the plant and this may result to low yield, reduction in pod size, poor yield quality, and loss in market value [6-8].

The damages caused by these field pests have been mainly controlled with organochlorines

(DDT, Thiodanetc); organophosphates (Monocrotophos, Dimethoate, etc); carbamates (Unden, Carbofuran, etc) and, more recently, pyrethroids (Permethrin, Cypermethrin, Imidacloprid, etc) [9-11].

Despite the efficacy of these synthetic insecticides, several adverse effects have been reportedly resulting from their misuse. These include; human poisonings, destruction of natural enemies, insecticide resistance, crop pollination problem due to honey bee losses, domestic animal poisonings, contaminated livestock products, fish and wildlife losses [12]; contamination of underground water and rivers, high persistence of the compounds, resurgence and genetic resistance of pests, adverse affect on non–target beneficial pests [13], etc. Cypermethrin 10 EC, used as a standard in this study, is a synthetic insecticide of choice in Nigeria. It is used in large scale commercial agricultural application as well as in consumer products for domestic purposes [14].

The ever- increasing problems associated with synthetic insecticides have synergized keen interest in the use of plant products as bioinsecticides. Botanicals are relatively safe, non-persistent, eco-friendlier and readily available [15]. Pyrethrum, a bioinsecticide, processed from the dried flower heads of *Chrysanthemum cinerariifolium* is said to have low mammalian toxicity, broad spectrum activity, is environmentally friendly and fast acting [16, 17]. Despite the worldwide acclaim for this plant, its cultivation and use in Nigeria is very limited.

Due to a dearth of information on the field pests of *A. hypogaea* in the Owerri west ecological zone of southeastern Nigeria, there is a compelling need to ascertain these field pests at different growth stages of the crop. In consideration of the debilitating effects of synthetic chemicals, it is also pertinent to screen biopesticides, especially pyrethrum, which would not be harmful to the farmers or end users of the treated produce.

## 2. MATERIALS AND METHODS

### 2.1 Site Location

The experiment was carried out at the Teaching and Research Farm of the Federal University of Technology, Owerri, Nigeria, from March to November, 2012. The area is between latitude 5°25' N and longitude 7°2' E in the Tropical Rainforest Zone of southeastern Nigeria. The experiment was rain-fed.

### 2.2 Collection and Preparation of Test Materials

Seeds of *Arachis hypogaea*, SAMNUT 23, were sourced from the Institute for Agricultural Research, Ahmadu Bello University, Zaria, Nigeria. This variety is also known as ICGV-IS 96894 under International Crops Research Centre for Semi-Arid Tropics (ICRISAT) nomenclature [18]. The variety is early maturing (90-100 days), has red seed colour with potential pod yield of 2,000kg/ha [19]. Pyrethrum, from the flower head of the plant, *Chrysanthemum cinerarifolium* Treviranus, was collected from a farm in Mawingo, Central Kenya. It was shade dried to prevent loss of active principle by sunlight, crushed into powdery form with pestle and mortar and weighed out at 0.25, 0.50 and 0.75 g. The weighed samples were mixed with 100 ml (respectively) of water in a container and allowed to settle for three hours. The solution was filtered through a fine muslin cloth and the active principle recovered and applied immediately.

Cypermethrin 10 E. C. was purchased from an agrochemical store in Owerri, Imo Sate, Nigeria. The insecticide was measured out with a syringe at 0.5, 1.0 and 1.5 ml, and mixed with 100 ml of water. Approximate recommended rate is 1.0 ml Cypermethrin/100 ml of water. The procedure was repeated on each spraying day.

### 2.3 Germination Test

Germination test was carried out by random selection of 30 wholesome seeds from the groundnut seeds kept in a bag. Ten seeds each were placed inside 3 Petri-dishes that contained absorbent Whatman's No. 44 filter papers moistened with water. Daily checks were carried out for 10 days and germination rate recorded from the 4th day.

### 2.4 Land Preparation

The site was cleared, pegged and beds made. Each bed had a dimension of 1m x 1m with furrows of 1 m in between. Altogether, the field contained a total of 24 beds.

### 2.5 Manure Application

Organic manure (cured poultry droppings) was basally applied on each of the beds and was incorporated into the soil at a rate of 2 kg per bed before planting.

## 2.6 Sowing and Weeding

The seeds were planted at a spacing of 30 cm x 15 cm. Two seeds were planted per hole, which was rogued down to a plant/hole after germination. This gave a total of 22.0 plants per bed and 220,000 plants per hectare.

Weeding was done at 3 WAP and later at intervals with the use of hoes and hand pulling (rogueing).

## 2.7 Field Application of Plant Extract and Cypermethrin 10 EC

The plant extracts and Cypermethrin 10 EC, in water, were foliar- sprayed using a 250 ml hand sprayer and were applied thrice at weekly intervals.

## 2.8 Pest Sampling

Insect pests were collected with a sweep net, cellophane bags and sample bottles. Others were hand-picked using hand gloves and plastic forceps. Samplings were carried out at 4 WAP (onset of flowering), 6 WAP (initiation of pegging, one to 2 weeks after fertilization) and 8 WAP (during pod development). These were done a day before and 2 days after each spray regime and twice during the pod and seed development stages. Collected insects were stored with chloroform and later identified in the laboratory.

## 2.9 Leaf Damage Assessment

Leaf damage assessment was done through visual recording of the number of leaves damaged by insects. It was carried out by counting the number of leaves damaged by insects on 6 selected and tagged stands. Damage assessments were recorded at 4, 6 and 8 weeks after planting (WAP).

## 2.10 Yield Measurement

Yield measurements were achieved through the following parameters:

### 2.10.1 Seed weight

The seed weight was measured after the pods from each bed were harvested and the seeds from dehisced pods weighed, using Camry Emperor weighing balance (model J1111427541) in kilogram (kg).

### 2.10.2 Shell weight

The shell weight measurement was carried out after drying under shade.

### 2.10.3 Pod density

The pod density was assessed twice. First, immediately after harvest, the pods from each bed were washed to remove soil and then weighed in the field. The second measurement was taken after drying, under room temperature in the laboratory.

## 2.11 Data Analysis

The data collected were subjected to Analysis of Variance (ANOVA) for Randomized Complete Block Design (RCBD) in a 2 x 4 factorial arrangement replicated three times. GENSAT Computer Software for data analysis was used and mean separation procedure was as described by [20] using Least Significant Difference at P = .05 level of significance.

## 3. RESULTS AND DISCUSSION

The major arthropod pests identified were termites, *Macrotermes bellicosus* Smeathman, 1781 (*Blattodea*: *Termitidae* -Macrotermitinae); millipedes, *Peridontopyge* spp. (*Diplopoda*: *Odontopygidae*); white grub larvae, *Helotrichia serrata* Fabricius, 1781 (*Coleoptera*: *Scarabaeidae*) and Nigerian grasshoppers, *Oedaleus nigeriensis* Uvarov, 1926 (*Orthoptera*: *Acrididae*) (Table 1). Other arthropod pest incidences were occasional and insignificant to report. The identified pests were in consonance with earlier works which implicated termites (*Odontotermes* spp.), white grubs (*Holotrichia consanguinea*, *H. serata*) and millipedes (*Peridontopyge* spp.) as the major, widespread, economic pests of groundnut in Sub-Saharan Africa [6-8]. Other coleopteran pests such as wireworms (*Agriotes lineatus*) and false wireworms (*Gerocephalum* spp.) have been reported to be of occasional importance [21].

The effect of pyrethrum, cypermethrin and their application rates on the field pests and leaf damage of groundnut plants at different growth stages are also as contained in Table 1. Statistical analysis of the main effect showed that cypermethrin and pyrethrum applications reduced total pest incidence (3.25, 2.50, 2.10 insects) and (4.09, 3.62, 3.42 insects), respectively, when compared with unsprayed plots (6.35, 6.16, 6.20 insects) at 4, 6 and 8 WAP. Except for *O. nigeriensis* (1.75 and 0.50) and *M. bellicosus* (0.50 and 1.25) at 4 and 8 WAP, respectively, insecticide type had no significant effect on the population of majority of sampled pests. Sprayed plots had less damaged leaves - 2.37, 2.73, 2.80 (average, 2.63) (cypermethrin) and 2.58, 2.90, 2.83 (average, 2.77)(pyrethrum) as against 3.17, 3.60, 4.52 (average, 3.76)(control) in unsprayed plots at 4, 6 and 8 WAP, respectively. Insecticide type also had no significant effect on leaf damage control.

The second rate gave the second best protection of groundnut against *M. bellicosus* (0.17, 0.50 and 0.50 insects) at 4, 6 and 8 WAP, respectively. The third (highest) rate recorded the lowest number (0.00, 0.17 and 0.50) of sampled *M. bellicosus* across the growth stages, respectively. Expectedly, the control plots were the most infested by the termites at all the stages. The various application rates followed the same efficacy trend against other sampled field pests.

Leaf damage result showed that the various rates of application significantly (P = .05) reduced the number of damaged leaves sampled across the growth stages. Leaves from groundnut plants sprayed with the third (highest) rate were the least damaged (1.82, 2.10 and 2.23 leaves) at 4, 6 and 8 WAP, respectively. Comparatively, the control plots recorded the highest (3.17, 3.60 and 4.52) number of damaged leaves, respectively. It could be reasonably inferred that cypermethrin and pyrethrum had basically the same statistical effect on pest incidence and leaf damage of the plant.

The interactive effect of application rate with insecticide type is shown in Table 2. There was only a significant interaction on the number of sampled *O. nigeriensis* all through the plant's growth stages. Applying cypermethrin at the third rate (1.50 ml/ 100 ml) recorded the lowest (0.33, 0.00 and 0.67) number of sampled target pests at 4, 6, and 8 WAP, respectively.

Table 3 records the effect of application rates and insecticide types on fresh pod, dry pod, seed and shell weights. The sprayed plots had increased fresh pod weight (0.61, 0.52 kg in cypermethrin and pyrethrum sprayed plots, respectively) as against 0.14 kg in unsprayed plots. Seed weights (0.26, 0.22 kg in cypermethrin and pyrethrum sprayed plots, respectively), were significantly distinct from the control (0.06 kg).There were no significant differences (P = .05) recorded in the dry pod or shell weights based on insecticide types.

There were significant (P= .05) differences with the various application rates on fresh pod and seed weights, respectively. The third rate gave the highest (0.57 and 0.30 kg) fresh pod and seed weights, respectively. The second rate was next with 0.56 and 0.27 kg fresh pod and seed weights, respectively. The efficacy of the insecticides was dose related as higher rates gave better performances.

Interaction effects of application rate with insecticide type on fresh pod, dry pod, seed and shell weights were not statistically significant (Table 4). However, the highest rate of cypermethrin (1.50 ml/ 100 ml) accounted for the highest fresh pod weight (0.58 kg), dry pod weight (0.47 kg), seed weight (0.32 kg) and shell weight (0.17 kg). These were, however, not statistically different from the effect of pyrethrum at the highest rate (0.75 g/ 100 ml).

Though cypermethrin performed marginally better than pyrethrum in some parameters, both treatments, however, showed no statistically significant differences (P = .05) in most parameters studied.

The effectiveness of cypermethrin in the control of insect pests of cotton, fruits, vegetable crops and as an indoor insecticide has been noted [22]. It has been reported that cypermethrin 10% EC, when sprayed at 10% w/v concentration, controlled the post flowering insect pests and increased pod and seed weight/plant of cowpea [23] and in conjuction with dimethoate gave the highest economic cowpea grain yield [24].

**Table 1. Main effects of application rate and insecticide type on target field pests of groundnut and leaf damage at different growth stages**

| Treatments | 4 WAP | | | | | | 6 WAP | | | | | | 8 WAP | | | | | |
|---|---|---|---|---|---|---|---|---|---|---|---|---|---|---|---|---|---|---|
| Application Rates | Mb | Ps | Hs | On | Total | LD | Mb | Ps | Hs | On | Total | LD | Mb | Ps | Hs | On | Total | LD |
| Control | 1.17 | 1.17 | 1.17 | 2.84 | 6.35 | 3.17 | 1.50 | 0.83 | 1.33 | 2.50 | 6.16 | 3.60 | 1.83 | 1.00 | 1.00 | 2.37 | 6.20 | 4.52 |
| First rate | 0.33 | 0.50 | 0.50 | 2.17 | 3.50 | 2.57 | 0.83 | 0.33 | 0.33 | 1.60 | 3.09 | 2.88 | 0.67 | 0.17 | 0.33 | 1.34 | 2.51 | 302 |
| Second rate | 0.17 | 0.50 | 0.50 | 1.84 | 3.01 | 2.33 | 0.50 | 0.33 | 0.50 | 0.84 | 2.17 | 2.67 | 0.50 | 0.00 | 0.30 | 0.67 | 1.47 | 2.77 |
| Third rate | 0.00 | 0.33 | 0.33 | 1.17 | 1.83 | 1.82 | 0.17 | 0.00 | 0.00 | 0.50 | 0.67 | 2.10 | 0.50 | 0.00 | 0.17 | 0.17 | 0.84 | 2.23 |
| LSD 0.05 | **0.816** | NS | NS | **0.204** | | **0.396** | **0.758** | NS | **0.532** | **0.304** | | **0.359** | **0.876** | NS | NS | **0.731** | | **0.764** |
| **Insecticide Type** | | | | | | | | | | | | | | | | | | |
| Cypermethrin | 0.42 | 0.58 | 0.50 | 1.75 | 3.25 | 2.37 | 0.50 | 0.33 | 0.42 | 1.17 | 2.50 | 2.73 | 0.50 | 0.25 | 0.42 | 0.93 | 2.10 | 2.80 |
| Pyrethrum | 0.42 | 0.67 | 0.75 | 2.25 | 4.09 | 2.58 | 1.25 | 0.42 | 0.67 | 1.55 | 3.62 | 2.90 | 1.25 | 0.33 | 0.50 | 1.34 | 3.42 | 2.83 |
| LSD 0.05 | NS | NS | NS | **0.123** | NS | NS | **0.619** | NS | NS | NS | NS | NS | NS | NS | NS | NS | NS | NS |

**Key:** NS: Non Significant; Mb: Macrotermes bellicosus (Termites); Ps: Peridontopyge spp (Millipedes)
Hs: Helotrichia serrata (White grub larvae); On: Oedaleus nigeriensis (Nigerian grasshoppers).
LD: Leaf damage

**Rates:** Pyrethrum Extract:    Rate 1 = 0.25 g/100 ml of water; Rate 2 = 0.50 g/100 ml of water;
Rate 3 = 0.75 g/100 ml of water;

Cypermethrin 10 EC:  Rate 1 = 0.50 Ml/100 Ml of Water; Rate 2 = 1.00 Ml/100 Ml Of Water;
Rate 3 = 1.50 Ml/100 Ml of Water

**Table 2. Interactive effects of application rate with insecticide type on target field pests of groundnut and leaf damage at different growth stage**

| Treatments | | 4 WAP | | | | | | 6 WAP | | | | | | 8 WAP | | | | | |
|---|---|---|---|---|---|---|---|---|---|---|---|---|---|---|---|---|---|---|---|
| Application Rates | Insecticide Type | Mb | Ps | Hs | On | Total | LD | Mb | Ps | Hs | On | Total | LD | Mb | Ps | Hs | On | Total | LD |
| Control | CYP | 1.33 | 1.33 | 1.00 | 2.67 | 6.33 | 3.10 | 1.33 | 0.67 | 1.33 | 2.33 | 5.66 | 3.50 | 1.00 | 1.00 | 1.00 | 2.40 | 5.40 | 4.30 |
| | PYM | 1.00 | 1.00 | 1.33 | 3.00 | 6.33 | 3.23 | 1.67 | 1.00 | 1.67 | 2.67 | 7.01 | 3.67 | 2.67 | 1.00 | 1.00 | 2.34 | 7.01 | 4.73 |
| First rate | CYP | 0.00 | 0.33 | 0.33 | 2.00 | 2.66 | 2.47 | 0.67 | 0.33 | 0.50 | 1.33 | 2.83 | 2.83 | 0.33 | 0.00 | 0.33 | 1.00 | 1.66 | 2.93 |
| | PYM | 0.33 | 0.67 | 0.67 | 2.33 | 4.00 | 2.67 | 1.00 | 0.33 | 0.67 | 1.87 | 3.87 | 2.97 | 1.00 | 0.33 | 0.33 | 1.67 | 2.33 | 3.10 |
| Second rate | CYP | 0.00 | 0.33 | 0.33 | 1.67 | 2.33 | 2.27 | 0.33 | 0.33 | 0.33 | 0.33 | 0.67 | 2.67 | 0.33 | 0.00 | 0.33 | 0.33 | 0.99 | 2.53 |
| | PYM | 0.33 | 0.67 | 0.67 | 2.00 | 3.67 | 2.40 | 0.67 | 0.33 | 0.33 | 1.00 | 2.33 | 2.73 | 0.67 | 0.00 | 0.33 | 1.00 | 2.00 | 3.00 |
| Third rate | CYP | 0.00 | 0.33 | 0.33 | 0.67 | 1.33 | 1.63 | 0.00 | 0.00 | 0.00 | 0.33 | 0.33 | 2.10 | 0.33 | 0.00 | 0.00 | 0.00 | 0.33 | 1.77 |
| | PYM | 0.00 | 0.33 | 0.33 | 1.67 | 2.33 | 2.00 | 0.33 | 0.00 | 0.00 | 0.67 | 1.00 | 2.23 | 0.67 | 0.00 | 0.33 | 0.33 | 1.33 | 2.70 |
| LSD 0.05 | | NS | NS | NS | 0.311 | | NS | NS | NS | NS | 0.424 | | NS | NS | NS | NS | 0.798 | | NS |

Key: CYP.: Cypermethrin; PYM.: Pyrethrum; NS: Non Significant; Mb: Macrotermes bellicosus (Termites); Ps: Peridontopyge spp (Millipedes); Hs: Helotrichia serrata (White grub larvae); On: Oedaleus nigeriensis (Nigerian grasshoppers); LD: Leaf damage

Rates: Pyrethrum Extract: **Rate 1** = 0.25 g/100 ml of water; **Rate 2** = 0.50 g/100 ml of water; **Rate 3** = 0.75 g/100 ml of water;

Cypermethrin 10 EC: **Rate 1** = 0.50 ml/100 ml of water; **Rate 2** = 1.00 ml/100 ml of water; **Rate 3** = 1.50 ml/100 ml of water

**Table 3. Main effects of application rate and insecticide type on fresh, dry, seed and shell weights (Kg)**

| Treatments | Fresh Pod Weight | Dry Pod Weight | Seed Weight | Shell Weight |
|---|---|---|---|---|
| **Application Rates** | | | | |
| Control | 0.14 | 0.31 | 0.06 | 0.11 |
| First rate | 0.53 | 0.43 | 0.23 | 0.12 |
| Second rate | 0.56 | 0.44 | 0.27 | 0.14 |
| Third rate | 0.57 | 0.46 | 0.30 | 0.15 |
| **LSD 0.05** | **0.876** | **NS** | **0.057** | **NS** |
| **Insecticide Type** | | | | |
| **Cypermethrin** | 0.61 | 0.42 | 0.26 | 0.14 |
| **Pyrethrum** | 0.52 | 0.40 | 0.22 | 0.12 |
| **LSD 0.05** | **0.079** | **NS** | **0.026** | **NS** |

*Key: NS: Non Significant*

**Table 4. Interactive effects of application rate with insecticide type on fresh, dry, seed and shell weights (Kg)**

| Treatments | | Fresh Pod Weight | Dry Pod Weight | Seed Weight | Shell Weight |
|---|---|---|---|---|---|
| **Application Rates** | **Insecticide Type** | | | | |
| Control | CYP | 0.48 | 0.32 | 0.17 | 0.12 |
| | PYM | 0.50 | 0.33 | 0.16 | 0.11 |
| First rate | CYP | 0.53 | 0.43 | 0.27 | 0.12 |
| | PYM | 0.53 | 0.42 | 0.20 | 0.12 |
| Second rate | CYP | 0.57 | 0.45 | 0.28 | 0.15 |
| | PYM | 0.55 | 0.43 | 0.25 | 0.12 |
| Third rate | CYP | 0.58 | 0.47 | 0.32 | 0.17 |
| | PYM | 0.57 | 0.45 | 0.28 | 0.13 |
| **LSD 0.05** | | **NS** | **NS** | **NS** | **NS** |

*Key: NS: Non Significant*

The efficacy of the insecticide could be as a result of its action as a stomach and contact poison and its debilitating action on the nervous system [25]. It has also been reported that due to its low vapour pressure ($1.3 \times 10^9$ mmHg at 20°C), the insecticide is absorbed as aerosols through stagnant surfaces like soil and foliage which are further exposed to atmospheric oxidation and solar radiation [26]. Though cypermethrin, which contain synthetic pyrethroids, is often dubbed as "safe as chrysanthemum flowers", it should be noted that they are chemically engineered to be more toxic with longer breakdown times, often formulated with synergists for increased potency, thereby compromising the human body's ability to detoxify the pesticide [27]. Cypermethrin could, therefore, cause debilitating health and environmental hazards [28, 29].

The performance of pyrethrum extracts in the experiment is commendable when related to the crude extraction method employed. Its action has been attributed to the presence of pyrethrin as an active insecticidal component [30]. Pyrethrin is reported to attack the nervous systems of all insects, and when not present in lethal doses, could act as a repellent and 'exciter' – increasing their activity and ability to 'flush' out insects from their hiding places thereby increasing their exposure with the insecticide [17].

Earlier report [31] has implicated six biologically active chemicals in pyrethrin: Pyrethrin I ($C_{21}H_{28}O_3$), Pyrethrin II ($C_{22}H_{28}O_5$), Cinerin I ($C_{20}H_{28}O_3$), Cinerin II ($C_{21}H_{28}O_5$), Jasmolin I ($C_{21}H_{30}O_3$) and Jasmolin II ($C_{22}H_{30}O_5$). Pyrethrin I, cinerin I, and jasmolin I are esters of chrysanthemic acid whereas pyrethrin II, cinerin II, and jasmolin II are esters of pyrethric acid. Chrysanthemic and pyrethric acids combine with one of three alcohols (pyrethrolone, cinerolone or jasmololone) to form the respective six active ingredients [32]. These acids are strongly lipophilic and rapidly penetrate many insects and paralyze their nervous system [33] and exert quick knockdown effects on a wide range of insect pests, causing paralysis within a few minutes and acting as contact poison that affects their central nervous system by blocking their nervous function [34].

Later findings by [35] showed that apart from pyrethrin, glandular trichomes are found in pyrethrum flower-head achenes and leaves. These trichomes are reportedly filled with many compounds among which sesquiterpene lactones (STLs) are the major constituent Pyrethrosin has earlier been established [36, 37] as the major isolate of STL which exhibits several biological properties. These isolates are insecticidal [38], cytotoxic [39], antibacterial [40], antifungal [41] and has root growth inhibitory activity [37]. They are also reported to have antifeedant activity against herbivores and are fungistatic against seedling-specific pyrethrum pathogens [35].

Pyrethrum STLs have, however been shown to cause allergic reactions [42]. This drawback has been ameliorated by improved refining techniques that yield pyrethrin oil preparations containing only trace amounts of STLs which no longer cause dermatitis [43].

Despite its efficacy, the natural pyrethrin insecticides have the desirable environmental properties of being both non-toxic to mammals and non-persistent [44].

## 4. CONCLUSION

The efficacy of pyrethrum is statistically comparable to cypermethrin in its ability to control the field insect pests of groundnut and could serve as alternative to synthetic pesticides in the management of these pests in southeastern Nigeria.

The use of the bioinsecticide could assist in arresting the prevailing dumping of thousands of tons of poisonous pesticides on agricultural soils. It is recommended that the potential of pyrethrum extract as a protectant for field crops and its cultivation in Nigeria be further explored.

Raising varieties and clones, with high pyrethrum content for the purpose, should be looked into, as the potency is dependent on content.

## ACKNOWLEDGEMENTS

The authors would wish to acknowledge the contribution of Mr. Garuba Adamu of the International Tobacco Company, Nigeria, who graciously procured, ex gratia, the pyrethrum for this experiment.

## COMPETING INTERESTS

Authors have declared that no competing interests exist.

## REFERENCES

1.   Harvest Choicein Science. Groundnut; 2010.
     Available:http://harvestchoice.org/commodities/groundnut

2.   FAO. Food and Agricultural Organisation of the United Nations. FAO Food outlook; 1990.
     Available:www.hort.purdue.edu/../peanut.html

3.   Himal J. Economic importance of groundnut oilseed crop. Oilseed Crop Groundnut India Meghmani Organics Limited, Ahmedabad. 2009a;14:12.
     Available:http://agropedia.iitk.ac.in/?q=content/economic-importance-groundnut

4.   Kenny A, Finn K. Groundnuts (*Arachis hypogea*), women and development: Impact on nutrition and women's roles in West Africa; 2004.
     Available:http://forest.mtu.edu/pcforestry/resources/studentprojects/groundnut.html
     Anonymous. Peanut as the world healthiest food; 2007.
     Available:http://www.en.wikipedia.org/wiki/peanut

5.   Johnson RA, Gumel MH. Termite damage and crop loss studies in Nigeria – the incidence of termite scarified groundnut pods and market samples. Tropical Pest Management. 1981;27: 343-350.

6.   Johnson RA, Lamb RW, Wood TG. Termite damage and crop loss studies – a survey of damage to groundnut. Tropical Pest Management. 1981;27: 325-342.

7.   Lynch RE, Ouedrago AP, Dicko I. Insect damage to groundnut in semi-arid tropical Africa. In Agrometeorology of Groundnut. Proceedings of an International Symposium, 21-26 August, ICRISAT Sahelian Centre, Niamey, Niger. Patancheru, A. P. 502 324, ICRISAT, India. 1986;175-183.

8.   Jat MK, Tetarwal AS. Pests of groundnut and its management. Krishisewa Agriculture Information Hub; 2014.
     Available:http://www.krishisewa.com/cms/disease-management/234-groundnut-pests.html

9.   Jasani H. Insect Pest Management in Groundnut. Agropedia; 2009.

Available:http://agropedia.iitk.ac.in/content/insect-pest-management groundnut

10. McGill NG, Bade GS, Vitelli RA, Allsopp PG. Imidacloprid can reduce the impact of the White grub *Antirogus parvule* on sugar cane. Crop Protection. 2003;22: 1169-176.

11. Pimentel D, Andow D, Dyson-Hudson K, Gallahan D, Jacobson S, Irish MK, Moss A, Schreiner1, Shepard M, Thompson T, Vinzant B. Environmental and Social Costs of Pesticides: A preliminary assessment. Oikos. 1980;34:126-140.

12. Grzywacz D, Leavett R. Biopesticides and their role in modern pest management in West Africa. Natural Research Institute/University of Greenwich Collaboration; 2012. Available:http://www.nri.org/news/archive/2012/20120413-biopesticides.htm

13. Ayoola SO, Ajani EK. Histopathological Effects of Cypermethrin on Juvenile African Catfish (*Clarias gariepinus*). World Journal of Biological Research. 2008;001(2)2:1-14.

14. Isman MB. Botanical insecticides, deterrents, and repellents in modern agriculture and an increasingly regulated world. Ann. Rev. Entomol. 2006;51:45–66.

15. Dayan FE, Cantrell CL, Duke SO. Natural products in crop protection. Bioorganic & Medicinal Chem. 2009; 17:4022–4034.

16. Pavela R. Effectiveness of some botanical insecticides against *Spodoptera Littoralis* Boisduvala (Lepidoptera: Noctudiae), *Myzus persicae* Sulzer (*Hemiptera: Aphididae*) and *Tetranychus urticae* Koch (*Acari*: *Tetranychidae*). Plant Protect. Sci. 2009;45:161-167.

17. Ndjeunga J, Ntare BR, Waliyar F, Echekwu CA, Kodio O, Kapran I, Diallo AT, Amadou A, Bissala HY, Da Sylva A. Early adoption of modern groundnut varieties in West Africa. Working Paper Series No. 24. Sahelian Centre, BP 12404 Niamey, Niger. International Crops Research Institute for the Semi-Arid Tropics. Patancheru, Andhra Pradesh, India. 2008;62. Available: http://oar.icrisat.org/2360/

18. ICRISAT. International Crops Research Institute for the Semi-Arid Tropics. The groundnut seed project. Development of sustainable groundnut seed systems in West Africa: Towards improving the productivity and quality of groundnut; 2003;1. Available:http://www.icrisat.org/gsp/default.asp

19. Wahua TAT. Applied statistics for scientific studies Africa-link press, Ibadan, Nigeria. 1999;108-248.

20. Umeh VC, Youm O and Waliyar F. Soil pests of Groundnut in Sub-saharan Africa .Intnl. J. Trop. Insect Sci. 2001; 21:23-32.

21. Takahashi N, Mikami M, Matsuda T, Miyamoto J. Photodegradation of the pyrethriod insecticide Cypermethrin in water and on soil surface. J. Pesticide Science. 1985;10:629-642.

22. Degri MM, Maina YT, RichardBI.Effect of Plant Extracts on Post Flowering Insect Pests and Grain Yield of Cowpea (*Vigna unguiculata* (L.) Walp.)) in Maiduguri, Semi Arid Zone of Nigeria. Journal of Biology, Agriculture and Healthcare. 2012;2(3):46-51.

23. Amatobi, CI. Insecticide application for economic production of cowpea grains in the Northern Sudan savannah of Nigeria. International Journal of Pest Management. 1995;41(1):14–18.

24. Jin H, Webster GRB. Persistence, penetration and surface availability of Cypermethrin and its major degradation products in elm bark. J. Agric. Food Chemistry. 1998;46:2851-2857.

25. Bacci E, Calamari D, Gaggiand C, Vighi M. An approach for the prediction of environment distribution and fate of Cypermethrin. Chemosphere. 1987;16:1373–1380.

26. Beyondpesticides. Pyrethroids/ Pyrethrins. Beyond Pesticides Rating: Toxic. 2014. Available:http://www.beyondpesticides.org/infoservices/pesticidefactsheets/toxic/pyrethroid.php

27. Elbetieha A, Daas SI, Khamas W, Darmani H. Evaluation of the toxic potential of cypermethrin pesticide on some reproductive and fertility parameters in the male rats. Arch. Environ. Contam. Toxicol. 2001;41:522-528.

28. ETN. Extension Toxicology Network. Cypermethrin: Pesticide Information Profiles; 1996. Available:http://ace.orst.edu/cgi-bin/mfs/01/pips/cypermet.htm

29. Morris SE, Davies NW, Brown PH, Groom T. Effect of drying conditions on pyrethrins content. Industrial Crop Production. 2006;23(1):9-14.

30. Moorman R, Nguyen K. Identification and qualification of the six active compounds in a pyrethrin standard. J. Assoc. Off. Anal. Chem. 1997;65:921-926.

31. Head SW. Composition of pyrethrum extract and analysis of pyrethrins. In: Pyrethrum: The Natural Insecticide. Cadisa JE (ed.). Academic Press. New York, NY. 1973;25-49.

32. Reigart JR, Roberts JR. *Recognition and Management of Pesticide Poisonings*, U.S. Environmental Protection Agency; 1999. 401 M Street SW (7506C), Washington, DC 20460. Available:http://www.epa.gov/pesticides/safety/healthcare/handbook/handbook.pdf

33. Bhat B, Menary R. Pyrethrum production in Australia: its past and present potential. J. Australian Institute of Agric. Sci. 1984;494:189-192.

34. Ramirez AM, Stoopen G, Menzel TR, Gols R, Bouwmeester HJ, Dicke M, Jongsma MA. Bidirectional secretions from glandular trichomes of pyrethrum enable immunization of seedlings. Plant Cell. 2012;10:4252–4265.

35. Doskotch RW, Elferaly FS. Isolation and characterization of (+)-sesamin and beta-cyclopyrethrosin from pyrethrum flowers. Canadian Journal of Chemistry. 1969;47:1139-1146.

36. Sashida Y, Nakata H, Shimomura H, Kagaya M. Sesquiterpene lactones from pyrethrum flowers. Phytochemistry. 1983;22:1219–1222.

37. Mourad LS, Salah OO, Osama S, Amal A. Insecticidal Effect of Chrysanthemum coronarium L. Flowers on the Pest Spodoptera littoralis Boisd and its Parasitoid Microplitis rufiventris Kok. with Identifying the Chemical Composition. Journal of Applied Sciences. 2008;8(10): 1859-1866.

38. Bach SM, Fortuna MA, Attarian R, de Trimarco JT, Catalán CA, Av-Gay Y, Bach H. Antibacterial and cytotoxic activities of the sesquiterpene lactones cnicin and onopordopicrin. Natural Product Communications. 2011;6(2): 163-166.

39. Picman AK, Towers GHN. Anti-bacterial activity of sesquiterpene lactones. Biochemical Systematics and Ecology. 1983;11:321–327.

40. Picman AK. Anti-fungal activity of helenin and isohelenin. Biochemical Systematics and Ecology. 1983;11:183–186.

41. Rickett FE, Tyszkiewicz K, Brown NC. Pyrethrum dermatitis. I. The allergenic properties of various extracts of pyrethrum flowers. Pesticide Science. 1972;3:57–66.

42. Ramirez AM, Saillard N, Yang T, Franssen MCR, Bouwmeester HJ, Maarten A. Jongsma MA. Biosynthesis of Sesquiterpene Lactones in Pyrethrum (*Tanacetum cinerariifolium*). PLoS ONE. 2013;8(5):e65030. DOI:10.1371/journal.pone.0065030.

43. Bullivant MJ, Pattenden G. Photodecomposition of natural pyrethrins and related compounds. Pesticide Science. 1976;7(3):231–35.

# 8

# Nitrogen (N) and Phosphorus (P) Fertilizer Application on Maize (*Zea mays* L.) Growth and Yield at Ado-Ekiti, South-West, Nigeria

Omotoso Solomon Olusegun[1*]

[1]*Department of Crop, Soil and Environmental Sciences, Ekiti State University, Ado-Ekiti, Nigeria.*

*Author's contribution*

*The sole author designed, analyzed and interpreted and prepared the manuscript.*

Editor(s):
(1) Edward Wilczewski, Faculty of Agriculture, University of Technology and Life Sciences in Bydgoszcz, Poland.
(2) Mintesinot Jiru, Department of Natural Sciences, Coppin State University, Baltimore, USA.
Reviewers:
(1) Chaminda Egodawatta, Department of Plant Sciences, Faculty of Agriculture, Rajarata University of Sri Lanka, Anuradhapura, Sri Lanka.
(2) Annet Namayanja, National Crops Resources Research Institute (NaCRRI)-Namulonge, Kampala, Uganda.
(3) Anonymous, Marathwada Krishi Vidyapeeth, Parbhani, Maharashtra, India.
(4) Anonymous, Debre Markos University, Ethiopia.

## ABSTRACT

Nutrient depletion as a result of continuous cultivation without supplementary addition of external inputs is a major challenge to agricultural productivity in South-west Nigeria. An experiment was set up to evaluate the effects of nitrogen (N) and phosphorus (P) fertilizer application rates on the performance of maize (*Zea mays* L.) in field trials at the Teaching and Research Farm, Ekiti State University, Ado-Ekiti, south-west Nigeria. The treatments consisted of 2 factors (i) N at 0, 30, 60, 90kg N.ha$^{-1}$ (ii) P at 0, 15, 30, 45kg P·ha$^{-1}$ in all possible combinations and laid out in a randomized complete block design arranged with three replicates. Data of plant height, leaf area, stem girth and cob length, cob diameter, 100 grain weight and grain yield were collected. The result showed that plant height, stem girth and leaf area· plant$^{-1}$ increased with N and P fertilizer rates. Cob length, cob diameter, 100 grain weight and grain yield, significantly (P=.05) increased with N and P application such that 90kg N·ha$^{-1}$ and 30kg N·ha$^{-1}$ gave the highest values. It may be concluded that application of the combination of N at 90kg·ha$^{-1}$ and P at 30kg·ha$^{-1}$ which produced the highest grain yield of maize could be regarded as the optimum for N and P in the study area. Therefore, further work should be carried out to ascertain the validity of this rate for maximum productivity.

*Corresponding author: E-mail: seguntoso@yahoo.com;

Keywords: Fertilizer; grain yield; maize; nitrogen; phosphorus.

# 1. INTRODUCTION

Cereals are the main crops grown in a wide range of environments in various parts of the world and can provide approximately 55% of proteins, 15% of fats, 70% of glucosides and 50-55% calories needed by man [1,2]. Maize is an important cereal crop in Nigeria, which is a useful ingredient in the formulation of livestock feed and other food preparations [3]. It is consumed fresh or processed, depending on the users' demand to provide raw materials for industrial preparation of several by-products like corn starch, corn oil, dextrose, corn syrup, corn flakes, cosmetics, wax, alcohol and tanning materials [4]. Thus, maize production and utilization are being promoted as viable ventures to farmers because of the numerous benefits and economic importance [5].

Maize is a heavy feeder of nutrients particularly nitrogen, phosphorus and potassium which are needed for good growth and high yield of crops [6]. The low fertility status of most tropical soils is a hindrance to production of maize, which removes a lot of nutrients from the soil. Thus, without an adequate nutrient supply, maize would fail to produce high grain yield [7]. Inorganic fertilizers supply the needed nutrients, and thus enhance plant luxuriant growth, development and yield [8].

Nitrogen is a major yield determining factor in maize production [9]. It is a critical component of protein, which controls the metabolic processes, required for optimum plant growth, and so must be available in sufficient quantity throughout the growing season of plant [10]. Unfortunately, the capacity of tropical soils to supply N declines rapidly as a result of agricultural activities, hence the inherent quantity of N derived from soil organic matter must be supplemented by other external sources.

Phosphorus is also required by maize for growth, being an essential component of nucleic acid, phosphorylated sugar, lipids and protein and so, plays a vital role in grain production [11]. It forms high-energy phosphate bonds with adenine, guanine and uridine, which act as carriers of energy for many biological reactions. It is important for seed and fruit formation and hastens crop maturation. Phosphorus hastens the ripening of fruits and can counteract the effects of excess N applied to the soil. Root growth and development are critical for early P uptake by plants, since it is relatively unavailable and immobile in many soils [12]. Thus, maize P requirement is high at the early stages of growth [13]. Adediran and Banjoko [14] reported positive and significant responses of maize to low P rates in the southern guinea savanna zones of Nigeria. Thus, in order to achieve high yields of maize, there is need for balanced N and P nutrition of maize. Several fertilizer trials, carried out in different locations in Nigeria showed that responses to applied N and P, in term of growth and grain yield depend on maize varieties with hybrid cultivars requiring higher fertilizer rates for optimum yield. Therefore, in view of the importance of maize as a grain crop and the role of nitrogen and phosphorus in its high productivity, this study was designed to examine the effect of varying nitrogen and phosphorus fertilizer rates on growth and yield of maize at Ado-Ekiti, Southwest Nigeria.

# 2. MATERIALS AND METHODS

## 2.1 Site Description

A field trial was conducted at the Teaching and Research Farm, Ekiti State University, Ado-Ekiti, Ekiti State, Nigeria, in 2010 and 2011 cropping seasons. The location experiences a tropical climate with distinct wet and dry seasons from late March to October and a dry spell in August. Mean annual rainfall, rainy days and temperature were 1367mm, 112 and 27°C, respectively. The soil of the study site has been identified as an Alfisols [15] derived from granitic rocks of the basement complex, highly leached with low to medium organic matter content. The site had been under continuous cultivation for some years to some arable crops, among which are maize, cassava, cocoyam before it was left to fallow for three years, prior to the commencement of this study.

## 2.2 Soil Sampling and Laboratory Analysis

Prior to planting, ten core samples randomly collected from 0-15cm soil depth, were bulked to form a composite sample, which was analyzed. The samples were thoroughly mixed, dried, crushed and sieved, using a 2 mm sieve. Particle size distribution was carried out by the hydrometer method [16]. The pH was determined in water (ratio1:1, soil: water). Soil pH in KCl (1:1), 20g of soil and 20ml of water was used and equilibrated for 30 minutes with occasional stirring. pH was determined and measured in KCl [17]. Organic carbon was determined by wet dichromate method [18] and the total N by the micro kjeldahl digestion method [19]. Available P was determined using Bray extraction method [20], while exchangeable cations were determine by extracting with neutral 1M $NH_4OAc$ in a solution ratio of 1:10 and measured by flame photometer. Magnesium was determined with an atomic absorption spectrophotometer. Effective cation exchange capacity (ECEC) was obtained using summation method (i.e. sum of Ca, K, Mg, Na and exchangeable acidity). The determination of exchangeable acidity was made using the extraction-titration method.

## 2.3 Treatments and Experimental Design

The experiment is a 2x4 factorial, the two factors are (i) N at 0, 30, 60, 90 kg N·$ha^{-1}$, (ii) P at 0, 15, 30, 45 kg P·$ha^{-1}$ and in all possible combination of N and P fertilizer laid out in a randomized complete block design arranged with three replicates. The source of N was urea (46% N) while that of P was single super phosphate (18% $P_2O_5$). A basal application of K (50kg $K_2O$·$ha^{-1}$) in form of muriate of potash (KCl) was applied to all treatment plots. Three seeds of early maturing maize variety were sown per hole at 75 x 25cm spacing in a 3x5m treatment plot and later thinned to one seedling per stand at one week after emergence giving a population of 53333 maize plants per hectare. Weeds were controlled manually by hand at 3 weeks after planting.

## 2.4 Data Collection

Data were collected on growth parameters of randomly selected maize plants in each plot. Plant height was measured from soil level up to the tip of highest leaf with a tape meter rule. Stem diameter was measured with the aid of Vernier Caliper from top middle and bottom portion of the same stem and the averages calculated. The leaf area per plant was calculated as the product of leaf length and widest middle portion of the leaf and then multiplied by a correction factor 0.75 [21]. Yield parameters included ear height, cob length, cob diameter, 100 seed weight and grain yield which was determined at 12.5% moisture content.

## 2.5 Data Analysis

Data collected on growth and yield parameters were subjected to analysis of variance (ANOVA) using Statistical Analysis System Institute Package [22]. The differences between treatment means were separated using Duncan's Multiple Range Test (DMRT) at 5% level of probability.

## 3. RESULTS AND DISCUSSION

### 3.1 Physical and Chemical Properties of the Soil Used

The data on physical and chemical properties of the soil prior to cropping are presented in Table 1. The data indicated that the soil were slightly acidic loamy sand. The values for N and P were below the critical level values obtained for soils in South-west Nigeria, as reported by [23]. The low N and P values can be explained in the light of the low organic matter content of the soil such that maize would be expected to benefit from added N and P fertilizers.

### 3.2 Effects of N and P Fertilizer Application on Growth Parameters of Maize

The result of this study showed that N and P individually had significant (P=0.05) effects on plant height. Plant height has been described as a measure of growth related to the efficiency in exploitation of environmental resources [4]. This growth character is directly linked with the productive potentials of plants in terms of fodder and grain yield [24]. The highest N dose in this study (Table 2) resulted in the tallest plant (139.8cm) which was not significantly different from 139.27cm obtained from 60kg N.$ha^{-1}$ application. Grazia et al. [25] noted that plant height can be increased with N application. Also, increase in plant height with nitrogen and phosphorus application has been reported by Ayub et al. [26] and Cheema [27]. This observation is also in agreement with the findings of Babatola [5] who reported that increasing level of fertilizer application led to increase in growth and yield of crops. Law-Ogbomo and Law-Ogbomo [28] reported that in *Zea mays*, plant height was increased with successive increment in NPK fertilizer application rate up to 600 kg·$ha^{-1}$. Stem girth increased significantly (P=.05) with N application to maximum values of 2.92cm while leaf area and number of leaves increased up till 60 kg.$ha^{-1}$ with values of 692$cm^2$ and 14.30 respectively after which there is reduction.

**Table 1. The physical and chemical properties of soil of the study site prior to cropping**

| Parameters | 2010 | 2011 | Methods |
|---|---|---|---|
| pH (1:1) $H_2O$ | 4.82 | 4.60 | Glass electrode pH meter [17] |
| Organic carbon (g·kg$^{-1}$) | 3.41 | 2.24 | Wet dichromate [18] |
| Total nitrogen (g·kg$^{-1}$) | 0.21 | 0.26 | Modified Kjeldal [19] |
| Available phosphorus (mg·kg$^{-1}$) | 4.06 | 5.60 | Bray 1[20] |
| **Exchangeable bases (cmol·kg$^{-1}$)** | | | |
| K | 0.17 | 0.20 | Flame photometer |
| Ca | 2.28 | 2.42 | Atomic absorption spectrophotometer |
| Mg | 0.92 | 0.45 | |
| Na | 0.43 | 0.60 | |
| Exch. Acidity | 0.17 | 0.16 | |
| Effective cation exchange capacity | 3.97 | 3.83 | |
| **Particle size analysis (g·kg$^{-1}$)** | | | |
| Sand | 786 | 813 | Hydrometer method [16] |
| Silt | 128 | 132 | |
| Clay | 140 | 54 | |
| Textural class | Loamy sand | Sandy loam | |

**Table 2. Effect of N and P fertilizers rate on growth parameters of maize**

| Treatments | Plant height (cm) | Stem girth (cm) | Leaf area (cm$^2$) | Number of leaves per plant |
|---|---|---|---|---|
| **Nitrogen (kg·ha$^{-1}$)** | | | | |
| 0 | 126.48c | 2.73b | 624.63a | 12.83b |
| 30 | 130.57b | 2.60c | 665.16a | 12.91b |
| 60 | 139.27a | 2.72b | 748.16a | 14.30a |
| 90 | 139.83a | 2.92a | 692.30a | 12.75b |
| **Phosphorus (kg·ha$^{-1}$)** | | | | |
| 0 | 110.79c | 2.40c | 588.98c | 12.58a |
| 15 | 135.27b | 2.65b | 672.58b | 12.83a |
| 30 | 135.22b | 2.80ab | 683.54b | 13.00a |
| 45 | 154.80a | 2.92a | 754.90a | 13.02a |
| Mean | 134.02 | 2.69 | 678.78 | 12.85 |
| SE± | 70.5 | 0.23 | 62.4 | 0.31 |

*Means followed by the same letter in each column are not significantly different at P=.05 using DMRT*

Phosphorus application increased maize plant height, stem girth, leaf area and number of leaves but the application rate of 15kg and 30kgP·ha$^{-1}$ in plant height, stem girth and leaf area was not different (Table 2). Plant height increased from 110.79cm in the control to 154.80cm at 45kg·ha$^{-1}$ P. The increase in leaf number was not significant with P application.

Ear height is strongly influenced by soil, water, nutrients, light situation and plant competition due to plant density whose optimization is necessary to maximize the genetic potential of maize cultivars. Ear height increased with N rates from 71.3cm in the control up till 81.0 cm at 60kg N·ha$^{-1}$ and decreased at 90kg N.ha$^{-1}$ (Table 3). However, this maximum ear height was not significantly different from 30kg N·ha$^{-1}$

rate. P application did not significantly increase ear height (Table 3).

## 3.3 Effects of N and P Fertilizer Application on Yield Components of Maize

Cob length and cob diameter increased significantly (P=.05) with N application. The maximum cob length (17.68cm) and cob diameter (3.92cm) were recorded at 90 kg N·ha$^{-1}$. The least cob length (15.19cm) and cob diameter (3.45cm) were recorded in control plot. However, maize fertilized with 30 kg N·ha$^{-1}$ and 60 kg N·ha$^{-1}$ rates were not significantly different in terms of these features. The significant response may probably be due to adequate supply of N which enhanced more photosynthetic activities of the plant. Crawford et al. [29] had stated that N is an essential requirement for cob and kernel growth in maize. Also, Akhtar and Silva [30] had observed significant increases in cob diameter and cob length with increase rates of nitrogen from different N sources.

Phosphorus application to maize was significantly affected, the highest value of cob length (17.60 cm) and cob diameter (4.73 cm) was obtained with 45kg P·ha$^{-1}$ which is 7.8 and 5.7% increase over the control. The maximum dry cob yield (12.61t·ha$^{-1}$) was obtained at 60kgN·ha$^{-1}$ treated plants whilst the control plant had the least (4.42t·ha$^{-1}$). The dry cob yields obtained with the application of 30, 60 and 90kg/ha N was106, 185 and 170% increase relative to the untreated plant respectively. Also for P application of 45kg·ha$^{-1}$ gave the maximum dry cob yield (13.4t·ha$^{-1}$) which was 72% increase over the untreated plants.

## 3.4 Effects of N and P Fertilizer Application on 100 Grain Weight of Maize

The effect of N and P fertilizer application on 100 grain weight is presented in Table 3. The 100 grain weight is an important yield determining factor; it expresses the magnitude of seed development for deriving the grain quality and yield per hectare [31]. The effects of N and P rates on 100  grain weight showed that N application resulted in significant (P=.05) increase as the rate increase and for P, it increased up to 30kg·ha$^{-1}$ after which there was reduction. This suggests a linear relationship between maize grain weight and increase rate of

N and quadratic relationship with P application. Ma et al. [32] had reported a linear increase in the weight of 100 grain in two varieties of maize and this was attributed to increase in nitrogen which plays very important roles in several physiological processes in plants.

## 3.5 Effects of N and P Fertilizer Application on Grain Yield of Maize

Grain yield is an important economical part of the plant which is available to man and livestock. It is affected by environmental factors and genetic potential of plant. Grain yield was significantly affected by both N and P application as indicated in Table 3. Sole application of N at 90kg·ha$^{-1}$ produced grain yield 3102.5 kg·ha$^{-1}$ compared to 1853.6 kg·ha$^{-1}$ in the control. Falhi [33] had reported that the use of nitrogen gives significant increases in maize grain yield. The increase in yield from application of N was mainly due to greater plant height, stem diameter and leaf area of the plant [34]. Nyende et al. [35] had stated that the rapid growth rate may have promoted photosynthetic activities in maize plants with resultant high carbohydrate production during fruiting, based on the premise that the growth of plant is positively correlated with their yields. Also, increase in fodder yield with N and P application has also been reported by [26,36,27]. Phosphorus fertilizer rate at 30 kg·ha$^{-1}$ produced grain yield of 3091.5 kg·ha$^{-1}$ and beyond which there was a significant reduction while the least value was recorded in the control plot. This could probably be as a result of plant growth which might be affected due to deficiency of Zn induced by high P levels as reported by Sinha et al. [37].

## Table 3. Effect of N and P fertilizers rate on yield and yield components of maize

| Treatments | Ear height (cm) | Cob length (cm) | Cob diameter (cm$^2$) | Dry cob weight (t·ha$^{-1}$) | 100 grain weight (g) | Grain yield (kg·ha$^{-1}$) |
|---|---|---|---|---|---|---|
| Nitrogen kg·ha$^{-1}$ | | | | | | |
| 0 | 71.26c | 15.19c | 3.45c | 4.42d | 20.45c | 1853.6d |
| 30 | 78.74a | 16.20b | 3.57b | 9.10c | 25.10b | 2403.2c |
| 60 | 80.98a | 16.73b | 3.60b | 12.61a | 26.30b | 2901.7b |
| 90 | 76.26b | 17.68a | 3.92a | 11.93b | 28.71a | 3102.5a |
| Phosphorus kg·ha$^{-1}$ | | | | | | |
| 0 | 75.19a | 16.32b | 3.01b | 3.64d | 19.34d | 1990.7d |
| 15 | 75.13a | 17.30a | 3.38b | 8.32c | 21.60c | 2521.4b |
| 30 | 79.17a | 17.50a | 4.32a | 11.45b | 24.21a | 3091.5a |
| 45 | 79.34a | 17.60a | 4.73a | 13.40a | 23.60b | 2106.3c |
| Mean | 77.21 | 17.18 | 3.86 | 9.20 | 22.19 | 2427.48 |
| SE± | 0.026 | 0.23 | 0.036 | 1.38 | 4.50 | 76.8 |

*Means followed by the same letter(s) in each column are not significantly different at P=.05 using DMRT*

### 3.6 Effects of N x P Fertilizer Application on Maize

Nitrogen and phosphorus and their interaction effect were significant (P=.05) on 100 grain weight and grain yield (Table 4). Since the interaction of N × P rate was significant; mean values for each treatment combination are presented in Table 4. Means Comparison of interaction effect indicated that among all interactions, the highest grain yield (3420 kg·ha$^{-1}$) was recorded in the combination of 90 kg N·ha$^{-1}$ with phosphorus application of 30 kg P·ha$^{-1}$ followed by the combinations of similar nitrogen level with lower rate of 90 kgN +15 kg P$_2$O$_5$·ha$^{-1}$. This combination produced 21% and 34% more yield than when N or P were applied alone and more than 78% the yield obtained from control plots. Irrespective of N rate, there was increase in 100 grain weight as P application increased up to 30 kg P except at the control plot where there is no nitrogen application.

**Table 4. Effects of N x P interactions on grain yield of maize**

| Fertilizer (kg·ha$^{-1}$) | | 100 grain weight(g) | Grain yield (kg·ha$^{-1}$) |
|---|---|---|---|
| Nitrogen | Phosphorus | | |
| 0 | 0 | 20.5f | 1921h |
| | 15 | 20.7f | 2321f |
| | 30 | 21.0f | 2534e |
| | 45 | 21.4f | 2106g |
| 30 | 0 | 23.2e | 2412ef |
| | 15 | 24.3d | 2607d |
| | 30 | 24.8d | 2532e |
| | 45 | 23.6e | 2710c |
| 60 | 0 | 25.6c | 2423ef |
| | 15 | 26.3c | 2650d |
| | 30 | 26.9b | 2734c |
| | 45 | 25.7c | 2801b |
| 90 | 0 | 26.9b | 2825b |
| | 15 | 27.6b | 3102b |
| | 30 | 28.8a | 3420a |
| | 45 | 28.5a | 2905bc |
| Mean | | 24.72 | 2625.19 |
| SE± | | 3.22 | 77.6 |

*Means followed by the same letter in each column are not significantly different at (P=.05) using DMRT*

### 4. CONCLUSION

The applications of N and P fertilizer thus have a profound significant influence on the growth and performance of maize. From this study, it may be concluded that application of the combination of N at 90kg·ha$^{-1}$ and P at 30kg·ha$^{-1}$ which produced the highest grain yield of maize could be regarded as the optimum for N and P in the study area. Therefore, further work should be carried out to ascertain the validity of this rate for maximum productivity.

### COMPETING INTERESTS

Author has declared that no competing interests exist.

### REFERENCES

1.  FAO, Food and Agriculture Organisation yearbook. 2007;60.

2.  Normohammadi G, Siadat A, Kashani A. Agronomy (Cereal), First Edn, Ahvaz, Iran, Shahid Chamran University publication. 2001;468.

3.  Akintoye HA, Olaniyan AB. Yield of sweet corn in response to fertilizer sources. Global Advanced Research Journal of Agricultural Science. 2012;1(5):110-116.

4.  Alimohammadi M, Mohsen Y, Peiman Z. Impact of nitrogen rates on growth and yield attributes of sweet corn grown under different Phosphorus levels. Journal of American Science. 2011;7(10):201-206.

5.  Babatola IA. Effects of NPK 15:15:15 fertilizer on the performance and storage life of okra (Abelmuschus esculentus). Proceedings of the Horticultural Society of Nigeria Conference. 2006;125-128.

6.  Onasanya RO, Aiyelari OP, Onasanya A, Oikeh S, Nwilene FE, Oyelakin OO. Growth and yield response of maize (zea mays) to different rates of nitrogen and phosphorus fertilizers in Southern Nigeria. World Journal of Agricultural Science. 2009;5(4):400-407.

7.  Adediran JA, Banjoko VA. Comparative effectiveness of some compost fertilizer formulations for maize in Nigeria. Nigerian Journal of Soil Science. 2003;13:42-48.

8.  Stefano P, Dris R, Rapparini F. Influence of growing conditions and yield and quality of cherry. II. Fruit J. Agric. And Envi. 2004;2:307-309.

9.  Reddy KJ. Nutrient stress. In: Physiology and molecular biology of stress tolerance in plants, Madhava Rao KV, Raghavendra AS, Janardhan Reddy K, (Eds.). Springer. 2006;187-217.

10. Lombin G. Evaluating the micronutrient fertility of Nigeria's semiarid savanna soils. 2. Zinc. Soil Sci.1983;136:42-47.

11. Khan HZ, Malik MA, Saleem MF. Effect of rate and source of organic materials on the production potential of spring maize (*Zea mays* L). Pak. J. Agric. Sci. 2008;45(1)40-43.

12. Mollier A, Pellerin S. Maize root system growth and development as influenced by phosphorus deficiency. Journal of Experimental Botany. 1999;50:487–497.

13. Ahn PM. Tropical soils & fertilizer use. TAP, Malaysia. 1993;128-130 &163-171.

14. Adediran JA, Banjoko VA. Response of Maize to Nitrogen, Phosphorus and Potassium fertilizers in the savanna zone of Nigeria. Commun. Soil Sci. Plant Anal. 1995;26:593-606.

15. Fasina AS, Aruleba JO, Omolayo FO, Omotoso SO, Shittu OS, Okusami T A. Properties and classification of five soils formed on granitic parent material in Ado-Ekiti, Southwestern Nigeria, Nigerian Journal of Soil Science. 2005;15(2):21–29.

16. Bouyoucos GJ. Hydrometer methods improved for making particle size analysis of soils. Soil Science Society of America proceeding. 1962;26:464-465

17. IITA selected methods of soil and plant analysis. International Institute of Tropical Agriculture Ibadan. Manual series. 1979;1.

18. Walkley A, Black IA. An examination of the Degtjareff method for determining soil organic matter and a proposed modification of the chronic acid titration method. Soil Science. 1934;37:29-39.

19. Bremmer JM, Mulraney CS. Nutrient Total In: Methods of soil analysis 2nd ed. AL, Page, et al. (Eds). ASA, SSA Medison Winsconsin. 1982;595-624

20. Bray RH, Kurtz LT. Determination of total organic and available forms of phosphorus in soils. Soil Sci. 1945;59:39-45.

21. Rajeshwari RS, Hebsur NS, Pradeep HM, Bharramagoudar TD. Effect of integrated nutrient management on growth and yield of maize. Karnataka J. Agric. Sci. 2007;20:399-400

22. SAS, Institute Lac, SAS/STAT User's Guide: Version 6, Fourth Edition, Carry, N C., SAS institute Inc. 2006;2:846.

23. Adeoye GO, Agboola AA. Critical level of soil plant available N, P, K, Zn, Mg, Cu and Mn on Maize leaf content in Sedimentary Soil of Southwestern Nigeria. Fert. Res. 1985;6:60-71.

24. Saeed IM, Abbasi R, Kazim M. Response of maize (*Zea mays*) to nitrogen and phosphorus fertilization under agro-climatic condition of Rawalokol, Azad Jammu and Kaslim and Kashmir, Pak. J. Biological Sci. 2001;4:949-952.

25. Grazia Jde, Tittonell PA, Germinara D, Chiesa A. Short communication: Phosphorus and nitrogen fertilization in sweet corn (*Zea mays* L. var. *saccharata* Bailey). Span J Agric Res. 2003;1(2):103-107.

26. Ayub M, Tanveer A, Ahmad R, Tariq M. Fodder yield and quality of two maize varieties at different nitrogen levels. Andhra Agric. J. 1997;47:7-11.

27. Cheema HNA. Yield and quality response of maize (*Zea mays* L.) fodder grown on different levels of phosphorus and seedling densities. M. Sc. Thesis, Department Agronomy University Agriculture Faisalabad Pakistan; 2000.

28. Law-Ogbomo K E, Law-Ogbomo JE. The Performance of *Zea mays* as Influenced by NPK Fertilizer Application. Notulae Scientia Biologicae. 2009;1(1):59-62.

29. Crawford TW, Rending VV, Broadbent FE. Sources; fluxes and sinks of nitrogen during early reproductive growth of maize (*Zea mays* L.). Plant Physiology. 1982;70:1654-1660.

30. Akhtar M, Silva JA. Agronomic traits and productivity of sweet corn affected by nitrogen and intercropping. Pak. J. Soil Sci. 1999;16:49-52.

31. Haseeb-Ur-Rehman, Asghar Ali, Muhammad Waseem Asif Tanveer, Muhammad Tahir, Muhammad Ather Nadeen, Muhammad Shahid Ibni Zamir. Impact of N application on growth and yield of maize (*Zea mays* L.) grown alone and in combination with cowpea (vigna *Unguiculata* L.). American-Eurasian Journal of Agric. And Environmental Science. 2010;7(1):43-47.

32. Ma BL, Dwyer LM, Gregorich EG. Soil nitrogen amendment effects on nitrogen uptake and grain yield of maize. Agron. J. 2004;9:650– 656.

33. Falhi GH. Crop growth and nutrient. Jahade Daneshgahi Publications. 1999;372.

34. Husnain MA. Effects of nitrogen application on fodder yield and quality of sorghum harvested at different times. M.Sc. Thesis, University Agriculture Faisalabad Pakistan; 2001.

35. Nyende P, Tenywa JO, Kidoido M. Weed profiles and management assessment for increase finger millet production in Uganda. African Crop Science Journal, 2001;9(3):507-516.

36. Ali S. Effects of different levels of nitrogen and seedling rates on growth, yield and quality of sorghum fodder. M.Sc. Thesis, Department of Agronomy University Agriculture Faisalabad, Pakistan; 2000.

37. Sinha RB, Sakal R, Kumar S. Sulphur and phosphorus nutrition of winter maize in calcareous soils. J. Indian Soc. Soil Sci. 1995;43(3):413-418.

# 9

# Influence of Host Plant Resistance and Disease Pressure on Spread of Cassava Brown Streak Disease in Uganda

K. Katono[1,2*], T. Alicai[2], Y. Baguma[2], R. Edema[1], A. Bua[2] and C. A. Omongo[2]

[1]College of Agricultural and Environmental Sciences, Department of Crop Production, Makerere University, P.O.Box 7062, Kampala, Uganda.
[2]National Crops Resources Research Institute (NaCRRI), Namulonge, Root Crops Programme, P.O.Box 7084, Kampala, Uganda.

***Authors' contributions***

*This work was carried out in collaboration between all authors. Author KK designed the study, wrote the protocol, performed the statistical analysis and wrote the first draft of the manuscript. Authors TA and RE reviewed the experimental design and all drafts of the manuscript. Authors YB, AB and CAO reviewed all drafts of the manuscript. All authors read and approved the final manuscript.*

Editor(s):
(1) Craig Ramsey, United States Department of Agriculture, Animal and Plant Health Inspection Service, Plant Protection and Quarantine, Center for Plant Health Science and Technology, 2301 Research Blvd., Suite 108, Fort Collins, CO 80526, USA.
(2) Anonymous.
Reviewers:
(1) Kathleen Hefferon, Cornell University, USA.
(2) Anonymous, Brazil.
(3) Njock Thomas Eku, Agronomic and Applied Molecular Sciences, University of Buea, Cameroon.

## ABSTRACT

Cassava brown streak disease (CBSD) is a major constraint to cassava production in Uganda. The disease is caused by two ipomovirus species: Cassava brown streak virus (CBSV) and Ugandan cassava brown streak virus (UCBSV), both transmitted by the whitefly vector (*Bemisia tabaci*). Since the outbreak of the CBSD epidemic in Uganda in 2004, knowledge of its spread in the field is still limited. In this study, five cassava genotypes with varying levels of resistance to CBSD: TME 204 (susceptible), I92/0067, MH 97/2961, MH 96/0686 (moderately tolerant) and NASE 3 (tolerant) were used to evaluate the effect of genotype and prevailing disease pressure on CBSD spread in Uganda. The experiment was established in a randomized Complete Block Design (RCBD) in three sites of varying CBSD disease pressure: high (Wakiso), moderate (Kamuli) and low (Lira) in

*Corresponding author: E-mail: kasifakats@yahoo.com;*

November, 2009 to November, 2010. Disease incidences (%), apparent infection rate (r), area under disease progress curves (AUDPC) were determined and population of the whitefly vector monitored monthly for 8 months. Genotype and disease pressure significantly affected CBSD incidence (P = .001), with Lira recording no noticeable disease spread even in the susceptible genotype TME 204. On the contrary, in Wakiso and Kamuli final disease incidence was maximum (100%) in the genotypes I92/0067, TME 204 and MH 97/2961 while the tolerant genotype NASE 3 had low final disease incidence of ≤ 5%. Mean whitefly population varied with site (P = .001) and there was a positive interaction between whitefly population and disease pressure hence the rapid CBSD spread in Kamuli and Wakiso. There was a high correlation (r = .994) between foliar and root CBSD incidence hence high CBSD root incidence in Kamuli and Wakiso. From these results, it is evident that high disease pressure, use of susceptible genotypes and high whitefly population significantly enhanced CBSD spread and development.

*Keywords: CBSD; disease pressure zones; whitefly; Bemisia tabaci; Uganda.*

## 1. INTRODUCTION

Cassava is a major subsistence crop in many parts of the world [1] and is a nourishing crop for resource-poor sub-Saharan African farmers [2,3]. In recent years, however, cassava production in Uganda and the coastal areas of East Africa has been constrained by Cassava Brown Streak Disease (CBSD) [4,5] which is caused by two ipomovirus species: Cassava brown streak virus (CBSV) and the Ugandan cassava brown streak virus (UCBSV) [5,6], both transmitted by the whitefly vector, *Bemisia tabaci* (Hemiptera; Aleyrodidae) [7].

Cassava brown streak disease epidemics spread fast while devastating large areas of cassava plantings and causing significant yield losses of between 70% - 100% [8]. Yield losses result from reduction of fresh weight and quality of the storage roots in susceptible varieties [8]. In Uganda, CBSD has spread to all major cassava producing districts, affecting most cassava genotypes including those that are highly resistant to cassava mosaic disease (CMD) [4]. Economic losses in affected areas are estimated at 30 million US dollars annually [9]. Cassava brown streak disease, although recognized on East African coastal areas since 1936 [10], is still among the most poorly understood diseases of cassava. Indeed, since the outbreak of the current CBSD epidemic in Uganda, there is limited knowledge of the factors that influence the spread of the disease. It is therefore important to carry out epidemiological studies to assess the effect of prevailing disease pressure and genotype on the incidence and spread of CBSD. Sound biological information, especially on varietal response was reported to be very critical in management of another cassava viral disease, CMD [11]. It was found that final incidence of CMD in a susceptible variety was dependent on the inoculum pressure in neighbouring fields [12].

Understanding the contribution of prevailing disease pressure, host tolerance and whitefly population dynamics to the general spread of disease will guide the development of appropriate area-specific disease control strategies as well as the development and deployment of CBSD resistant varieties which would contribute to the management of CBSD and its effects. This will ultimately contribute to securing the livelihoods of rural communities that primarily depend on cassava.

## 2. MATERIALS AND METHODS

### 2.1 Site, Genotypes and Experimental Design

The experiment was set up in November 2009 at three sites within Uganda; Wakiso (Namulonge) district in Central region which is at an elevation of 1200 m above sea level (asl) with a bi-modal rainfall pattern, Kamuli (Nabwigulu) district in the Eastern region which is at an elevation of 1100 m asl with a bi-modal rainfall pattern and Lira (Ngetta) district in Northern Uganda which is at an elevation of 1080 m asl with a uni-modal rainfall pattern. Previous surveys conducted by [13,14] showed that the three sites had high, moderate and low CBSD prevalence, respectively. Five commonly grown cassava genotypes varying in resistance to CBSD were used: TME 204 (susceptible), I92/0067, MH 97/2961 and MH 96/0686 (moderately tolerant), NASE 3 (tolerant). Clean planting materials of each of these genotypes were sourced from CBSD-free fields on basis of visual inspection in areas with no or low CBSD prevalence (Arua,

Oyam and Lira) and planted at each of the three sites. Absence of CBSV was further comfirmed by testing leaf samples using CBSV virus specific primers as described by [15]. The experiment was laid out in a randomized complete block design (RCBD) with four replicates, with plot sizes of 9 m x 9 m (10 plants x 10 plants i.e. 100 plants) and plant spacing of 1 m x 1 m with an alley of 1 m left between plots and 2 m between blocks. Weeds were controlled manually by hand hoeing monthly for the first 5 months, and thereafter weeding was done whenever necessary.

## 2.2 Data Collection and Analysis

Data on CBSD severity and incidence was collected at monthly intervals for twelve months starting a month after planting (MAP). The mean CBSD incidence was determined by expressing the number of plants showing CBSD foliar symptoms as a percentage of the total number of plants in a plot. Severity of CBSD infection was assessed using a scale of 1-5 where: 1 means no apparent symptoms, 2 means slight leaf chlorosis, 3 means severe leaf chlorosis and mild stem lesions, 4 means severe leaf chlorosis and severe stem lesions while 5 means defoliation, severe stem lesions and dieback [16]. Adult whiteflies were counted on the underside of the top five fully expanded leaves of the tallest shoot on each of the 15 randomly selected plants per plot starting 1 MAP for 8 months and mean numbers per plot were computed. Data on incidence and severity of CBSD on roots was collected at harvest (12 MAP). Thirty plants were harvested per plot and data was taken on total root weight, CBSD root incidence and CBSD root severity. Cassava brown streak disease root severity was assessed by slicing each root five times transversely for all the 30 plants and scoring using a scale of 1-5 where: 1 means no apparent necrosis, 2 means < 5% of the root is necrotic, 3 means 5% -10% of the root is necrotic, 4 means 10% – 25% of the root is necrotic and mild root constriction and 5 means >25% of the roots necrotic and severe root constriction.

Data on CBSD incidence, severity and adult whitefly population were first transformed for normality and then subjected to analysis of variance (ANOVA) using Genstat computer package 5 Release 3.2. Means were separated using the Least Significant Difference (L.S.D) test at 5% probability level. Actual disease progress (incidence %) curves (based on obviously diseased plants at each time of assessment),

were plotted to determine temporal spread of CBSD for each genotype and site. CBSD incidence was used for comparing the effect of prevailing disease pressure on the spread of CBSD. Symptom severity curves were also fitted for different genotypes.

The area under disease progress curve (AUDPC) was calculated using % incidence as described below [17]:

$$AUDPC = \sum_{i=1}^{(N-1)} [(X_{i+1} + X_i)/2][t_{i+1} - t_i]$$

- Where $\Sigma$ = summation; $X_i$ = disease incidence at time $t_i$ and $X_{i+1}$ = disease incidence at time $t_{i+1}$.

Apparent infection rates ($r$) of CBSD were calculated for each variety per location as described by [14] as follows:

$$r = (X_2 - X_1)/(t_2 - t_1)$$

- Where: $r$ is the apparent infection rate, $t_1$ is the time (months) of the first measurement, $t_2$ is the time of the second measurement, $x_1$ is the proportion of infection measured at time $t_1$ and $x_2$ is the proportion of infection measured at time $t_2$.

## 3. RESULTS

### 3.1 Progress of CBSD Foliar Incidence among Cassava Genotypes at Three Sites

Cumulative CBSD incidence varied significantly ($P = .001$) by both genotype and site. In Kamuli and Wakiso, final CBSD incidence at 12 MAP was low (≤5%) in NASE 3 and MH96/0686, and maximum in TME 204, MH97/2961 and I92/0067 (Table 1). There was no CBSD spread at all in Lira (Table 1).

### 3.2 Area under Disease Progress Curve (Audpc)

At Kamuli and Wakiso, AUDPC values for different genotypes varied significantly ($P = .05$). There was early infection of I92/0067, TME 204 and MH 97/2961 and CBSD incidence peaked at 7 to 10 MAP in Kamuli and Wakiso (Fig. 1). These genotypes had very high AUDPC values ranging from 1458 in MH 97/2961 to 4243.5 in TME 204 (Table 2). However, CBSD symptoms appeared late on NASE 3 and MM 96/0686, with no or low final CBSD incidence hence small or zero AUDPC values (Table 2, Fig. 1). Disease development was highest in Kamuli, followed by Wakiso.

**Table 1. Cassava brown streak disease incidence on cassava genotypes at three sites in Uganda**

| Genotype | Site | | |
|---|---|---|---|
| | Kamuli | Wakiso | Lira |
| I92/0067 | 100 | 100 | 0 |
| MH 97/2961 | 100 | 100 | 0 |
| MM96/0686 | 2.7 | 5 | 0 |
| NASE 3 | 2.5 | 0 | 0 |
| TME 204 | 100 | 100 | 0 |
| Mean | 61.1 | 61 | 0 |
| Lsd$^{0.05}$ | 2.9 | 4.3 | 0 |

**Table 2. Area under disease progress curve (AUDPC) for Cassava brown streak disease on cassava genotypes at three sites in Uganda**

| Genotype | Site | | |
|---|---|---|---|
| | Kamuli | Wakiso | Lira |
| I92/0067 | 3847.5 | 3087 | 0 |
| MH 97/2961 | 3289.5 | 1458 | 0 |
| MM 96/0686 | 0 | 0 | 0 |
| NASE 3 | 36 | 0 | 0 |
| TME 204 | 4243.5 | 3474 | 0 |
| Mean | 2283.3 | 1603.8 | 0 |
| Std Dev | 2096 | 1964 | ns |

## 3.3 Rate of Infection (r)

The rate of CBSD progress ($r$) varied significantly ($P = .05$) among genotypes. The rate of CBSD development over time among genotypes I92/0067, TME 204 and MH 97/2961 was high in the first six months but declined thereafter. By this time, CBSD incidence had almost reached maximum at both sites. I92/0067 and TME 204 had very high infection rates ranging from 0.7 to 0.9 at both sites. On the contrary, there was little or no infection in MM 96/0686 and NASE 3 (Table 3). Infection rate, was generally higher in Kamuli compared to Wakiso (0.0 – 0.9) (Table 3).

## 3.4 Progress of CBSD Severity on Cassava Genotypes

CBSD symptom severity varied significantly with genotype, site and crop age ($P = .001$). Final CBSD severity at 12 MAP was generally high at both Wakiso and Kamuli (Fig. 2). At both sites,

NASE 3 recorded the lowest CBSD severity sore of 2.5 (Kamuli) and 1 (Wakiso) (Fig. 2).

**Table 3. Apparent infection rate (r) for cassava brown streak disease on cassava genotypes at three sites in Uganda**

| Genotype | Site | | |
|---|---|---|---|
| | Kamuli | Wakiso | Lira |
| I92/0067 | 0.9 | 0.7 | 0 |
| MH 97/2961 | 0.8 | 0.3 | 0 |
| MM 96/0686 | 0 | 0 | 0 |
| NASE 3 | 0 | 0 | 0 |
| TME 204 | 0.9 | 0.7 | 0 |
| Mean | 0.5 | 0.3 | 0 |
| Std Dev | 0.5 | 0.4 | 0 |

*Disease incidence data for 3-6 MAP used*

## 3.5 Temporal Changes in Adult Whitefly Population

Mean whitefly populations varied with crop age and site ($P = .001$) but not with genotype. Colonisation of the crop by the whitefly vector was highest in younger plants, with peak infestation at 2 - 4 MAP (Fig. 3). In general, whitefly infestation varied over time and the decline in population occurred from 4 MAP (Fig. 3). At 2 MAP, high whitefly population was recorded in Kamuli with MH 97/2961 having highest mean number of 199.7 adults while MM 96/0686 had the lowest number (120.7). In Wakiso, whitefly population was generally low but increased steadily to peak at 4 MAP (Fig. 3): where the highest number was recorded on MM 96/0686 (163.5) and least in MH 97/2961 (89.1). Overall, the lowest whitefly populations were recorded in Lira with TME 204 and MM 96/0686 having the highest (10.9) and lowest (6.7) whitefly populations, respectively. However, the population of whiteflies suddenly peaked at 4 MAP and thereafter dropped drastically at six months (Fig. 3).

## 3.6 Relationship between whitefly population and CBSD incidence

There was an indirect relationship between whitefly population and CBSD incidence, with CBSD incidence increasing a month after an increase in whitefly population (Fig. 4). No relationship was observed for genotypes NASE 3 and MM96/0686 since disease symptoms were observed late when whitefly populations had dropped.

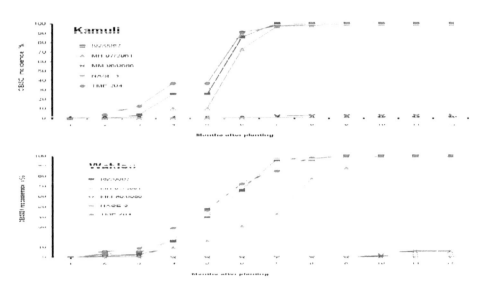

**Fig. 1. Disease progress curves for spread of cassava brown streak disease on cassava genotypes at three sites in Uganda**

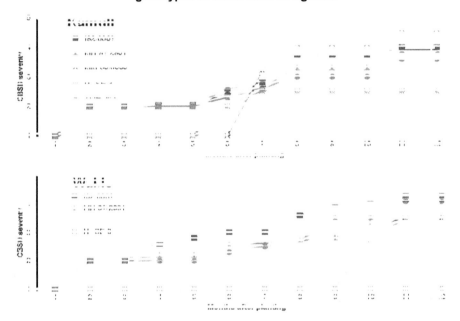

**Fig. 2. Cassava brown streak disease severity curves on cassava genotypes at three sites in Uganda**

## 3.7 Effect of CBSD on Yield of Cassava Tubers

Significant differences ($P = .001$) occurred on CBSD root incidence, severity and root weight among the different genotypes at the three sites. There was a high correlation ($r = 0.994$) between foliar and root CBSD incidence. High root incidence was recorded for all genotypes in Kamuli and Wakiso while Lira had very low incidence e.g. root incidence for TME 204 was 100% in Kamuli, 98.1% in Wakiso and 3.8% in Lira (Table 4). Also the genotypes MM96/0686 and NASE 3 which had low foliar incidence recorded the lowest root incidence (<20%) at all sites. CBSD root severity followed a similar trend as the root incidence: highest in Kamuli and Wakiso in TME 204, MH97/2961 and I92/0067. Among all genotypes, root weights were highest in Lira (Table 4). Among sites, inspite of the high foliar and root CBSD severity and incidence, I92/0067 had the highest total root weight.

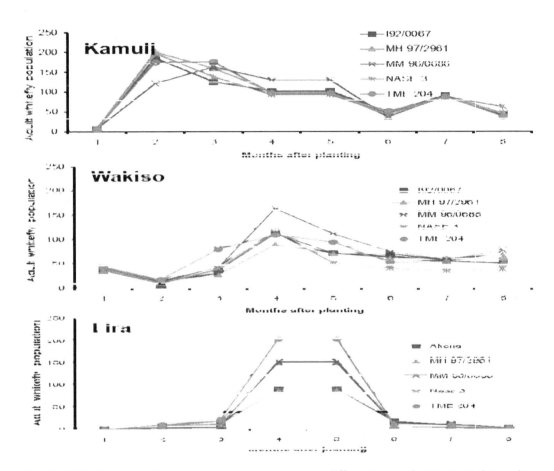

**Fig. 3. Whitefly population on cassava genotypes at different growth stages at three sites in Uganda**

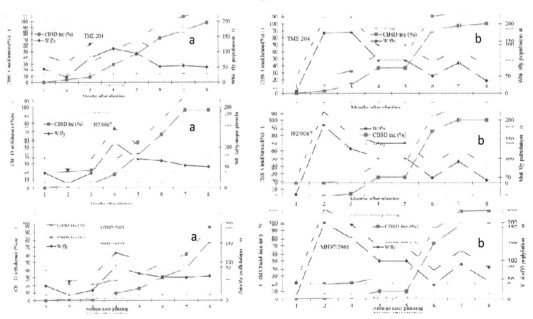

**Fig. 4. Relationship between cassava brown streak disease incidence (%) and whitefly population in Wakiso (a) and Kamuli (b) Uganda**

**Table 4. Relationship between foliar and root CBSD incidence and severity on cassava genotypes grown at three sites in Uganda**

| Genotype | Site | CBSD incidence (%) | | CBSD | Total root |
|---|---|---|---|---|---|
| | | Foliar | Root | Mean severity | Weight (Kgs) |
| I92/0067 | Kamuli | 100 | 93.6 | 3.6 | 105.7 |
| | Wakiso | 100 | 94.2 | 3.8 | 82.6 |
| | Lira | 0 | 0.7 | 3 | 152 |
| MH 97/2961 | Kamuli | 100 | 98.6 | 4.7 | 43 |
| | Wakiso | 100 | 86.9 | 4.6 | 63.3 |
| | Lira | 0 | 2.9 | 2.7 | 145.1 |
| MM 96/0686 | Kamuli | 2.7 | 14.9 | 3.7 | 94.3 |
| | Wakiso | 5 | 10.2 | 3.1 | 51 |
| | Lira | 0 | 0.8 | 2.3 | 139.5 |
| NASE 3 | Kamuli | 2.5 | 18 | 3.8 | 29.1 |
| | Wakiso | 0 | 5.5 | 2.3 | 13 |
| | Lira | 0 | 3.3 | 2 | 54.9 |
| TME 204 | Kamuli | 100 | 100 | 4.8 | 77.4 |
| | Wakiso | 100 | 98.1 | 4.6 | 56.8 |
| | Lira | 0 | 3.8 | 3.1 | 119.3 |
| Grand mean | | 40.7 | 42.1 | 3.3 | 81.8 |
| Lsd$_{0.05}$ | | 1 | 5.5 | 0.5 | 21.1 |

## 4. DISCUSSION

The variation in disease status indicated by the various parameters among the cassava genotypes demonstrated marked differences in tolerance to CBSD infection. Disease progressed more rapidly in the genotypes: I92/0067, TME 204 and MH 97/2961 which are apparently susceptible and moderately tolerant to CBSD. This revealed the significant role of susceptibility or resistance of cassava genotypes in influencing CBSD spread and development. These results agreed with the findings of [18,19] who observed that genotype susceptibility was an important factor in CMD spread. The genotypes MM96/0686 and NASE 3 recorded very low CBSD infection in both Kamuli and Wakiso, indicating their tolerance to CBSD over a wide range of environments.

There was no spread in Lira meaning that in areas where disease pressure is low, even the highly susceptible varieties can be deployed as long as virus-free planting material is used. On the contrary, where CBSD prevalence is high rapid spread of CBSD occurs, especially in the susceptible and moderately tolerant varieties. Similar findings were reported by [20] who showed that prevailing disease pressure in an area significantly influenced cassava mosaic disease (CMD) spread hence, differences in disease spread in different areas.

The results also showed differences in the whitefly population among the different sites, with Kamuli having the highest whitefly population followed by rapid spread of the disease especially in the susceptible and moderately tolerant varieties. In Lira where whitefly population was low, there was no disease spread. These findings confirmed the importance of both prevailing disease pressure and whitefly number in the spread of CBSD and are consistent with those of [21,22] who found that CMD, another whitefly viral-transmitted disease spread more rapidly in the high pressure zone where whitefly population is high compared to the low disease pressure zone.

Whiteflies infestation was highest on MM96/0686 (Kamuli), NASE 3 and MH97/2961 (Wakiso). However, the results showed that MM96/0686 and NASE 3 had the lowest infection rates and disease incidence despite the high vector population. In all sites, the results suggested that the tolerance to CBSD in these varieties was not due to resistance to the vector but rather the inherent genetic capacity of the varieties to suppress the virus. This observation was in agreement with the findings of [23,24].

Adult whiteflies occurred on cassava throughout the observation period but numbers were closely related to crop age. Low initial population may be due to the fact that young establishing plants did not attract whiteflies while the subsequent rapid vegetative growth produced large succulent leaves which were probably preferred by the whiteflies [25]. At later growth stages, the leaves senesced prompting the whiteflies to search for new growth for both oviposition and feeding [25].

The lack of a direct relationship between whitefly number and CBSD incidence at the time of this study was similar to findings by [26] in Cote d'Ivoire, who observed that the spread of CMD was not directly related to whitefly number. The increase in whitefly number and CBSD incidence one month after is consistent with the findings of [27] that CBSD symptoms on inoculated plants appeared after 26 -60 days. This period accounted for the latent period between CBSV infection and manifestation of the first CBSD symptoms.

Total root weight was generally lower in the high and moderate disease pressure zones especially in the highly susceptible TME 204. This was probably because the severe necrosis retarded root fill [28]. Although I92/0067 had high severity scores, total root weight was high suggesting that this genotype had a good degree of tolerance to CBSD. However, the high root incidence coupled with high CBSD root severity of I92/0067, MH97/2961 and TME 204 indicated severe loss of quality and production [8,29], hence unsuitability of these genotypes for use in high disease pressure zones. However, if they were deployed in areas where CBSD was endemic, they should be harvested at 8 to 9 MAP before severe root necrosis occurred [8,29].

The differences in the infection rate among different genotypes and across sites implied that different control measures may be needed for each site. Where disease pressure is low, sanitation procedures alone may be adequate in the management of CBSD, while in high disease pressure areas, the use of tolerant varieties is a prerequisite [30] but should be augmented by application of appropriate sanitation measures. These results agreed with earlier findings [22, 31] on the deployment of phytosanitation and plant resistance in management of CMD in low and high disease pressure areas.

## 5. CONCLUSION AND RECOMMENDATIONS

This study was set out to determine the effect of disease pressure, level of host plant resistance and whitefly vector population on the spread of CBSD. Based on our findings, we concluded that prevailing disease pressure and varietal resistance were the key factors in the spread of CBSD. The vector is slightly less important in CBSD spread [30] because of the semi-persistent nature of CBSV transmission [32]. The whitefly vector population was important in the dissemination of the virus provided a susceptible cassava genotype was grown, a ready source of the inoculum was available and that suitable environmental conditions prevailed.

MM96/0686 and NASE 3 showed the highest degree of tolerance to CBSD, however they were not popular among farmers. It is therefore recommended that these varieties be promoted and distributed among farmers in the CBSD hot-spot areas in order to reduce losses due to CBSD. Also the resistance genes in these genotypes could be useful in improving some of the commonly grown varieties such as I92/0067 which, despite showing high CBSD foliar and root severity gave good yields. However, these two tolerant varieties should be should be investigated at differential initial inocula to further validate their resistance status.

## ACKNOWLEDGEMENTS

This work was conducted with financial support from MSI- Uganda (Grant No. MSI / WAI / 02/08). We also acknowledge the technical and logistical support from the Root Crops Programme at National Crops Resources Research Institute (NaCRRI) and Makerere University. Dr. P. Paparu, Dr. R. Kawuki, Mr. I. Ramathani, Mr. G.A Okao, Mr. A. Abaca for their assistance during preparation of the manuscript

## COMPETING INTERESTS

Authors have declared that no competing interests exist.

## REFERENCES

1. Arbashi MM, Mohammed IU, Wasswa P, Hillocks RJ, Holt J, Legg JP, Seal SE, Maruthi MN. Optimisation of diagnostic RT-PCR protocols and sampling procedures for the reliable and cost-effective detection of Cassava brown streak virus. Journal of Virological Methods. 2010;163:353-359.

2. Baguma Y, Sun C, Ahlandsberg S, Mutisya J, Palmqvist S, Rubaihayo P, Magambo M, Egwang T, Larson H, Christer, J. Expression patterns of the gene encoding starch branching enzyme II

in the storage roots of cassava (*Manihot esculenta*). Plant Science. 2003;164:833-839.

3. Benesi M. Characterization of Malawian Cassava Germplasm for diversity, Starch extraction and it's Native and modified properties, PhD thesis, Free State University, SA; 2005.

4. Alicai T, Omongo CA, Maruthi MN, Hillocks RJ, Baguma Y, Kawuki R, Bua A, Otim-Nape GW, Colvin J. Re-emergence of cassava brown streak disease in Uganda. Plant Disease. 2007;91:24-29.

5. Winter S, Koerbler M, Stein B, Pietruszka A, Paape M, Butgereitt A. Analysis of cassava brown streak viruses reveals the presence of distinct virus species causing cassava brown streak disease in East Africa. Journal of General Virology. 2010;91:1365–1372.

6. Mbanzibwa DR, Tian YP, Tugume AK, Mukasa SB, Tairo, F, Kyamanywa S, Kullaya A, Valkonen JPT. Genetically distinct strains of cassava brown streak virus in the Lake Victoria basin and the Indian Ocean coastal area of east Africa. Archives of virology. 2009;154:353–359.

7. Maruthi MN, Hillocks RJ, Mtunda K, Raya MD, Muhanna M, Kiozia K, Rekha AR, Colvin J, Thresh JM. Transmission of cassava brown streak virus by *Bemisia tabaci* (Gennnadius). Journal of Phytopatology. 2005;153:307-312.

8. Hillocks RJ, Raya MD, Mtunda K and Kiozia H. Effects of brown streak virus disease on yield and quality of cassava in Tanzania. Journal of Phytopathology. 2001;149:389-394

9. Alicai T. Biotechnology applications to unravel and combat the ravaging cassava brown streak disease epidemic. Report of the Sixteenth Open Forum on Agricultural Biotechnology in Africa (OFAB) - Uganda Chapter. 2. July. Kampala; 2010.

10. Storey HH. Virus diseases of East African plants. VI. A progress report on studies of diseases of cassava. East African Agricultural and Forestry Journal. 1936;2:34-39.

11. Egesi CN, Ogbe FO, Akoroda M, Ilona P, Dixon A. Resistance profile of improved cassava germplasm to cassava mosaic disease in Nigeria. Euphytica. 2007;155: 215–224.

12. Legg JP, James B, Cudjoe A, Saizonou S, Gbaguidi B, Ogbe F, Ntonifor N, Ogwal H, Thresh JM, Hughes J. A regional collaborative approach to the study of CMD epidemiology in Sub-Saharan Africa. African Crop Science Conference Proceedings. 1997;3:1021-1033.

13. IITA, 2007/2008 (unpublished). Cassava pests and disease monitoring survey in Uganda.

14. NARO. Cassava pests and disease monitoring survey in Uganda; 2008.

15. Mbanzibwa DR, Tian YP, Tugume AK, Mukasa SB, Tairo F, Kamanywa S, Kullaya A, Valkonen JPT. Simultaneous virus-specific detection of the two cassava brown streak-associated viruses by RT-PCR reveals wide distribution in East Africa, mixed infections and infection in *Manihot glaziovii*. Journal of Virological Methods. 2011;171(2011):394-400.

16. Gondwe FMT, Mahungu NM, Hillocks RJ, Raya MD, Moyo CC, Soko MM, Chipungu F, Benesi IRM. Economic losses experienced by small-scale farmers in Malawi due to cassava brown streak virus streak virus disease. In: Cassava Brown Streak Virus Disease: Past, Present and Future. J. P Legg and RJ Hillocks, eds. Proc. Int. Workshop, Mombasa, Kenya, 27-30 October 2002. Natural resources international limited, aylesford, Uk. 2003;28-35.

17. Campbell CL, Madden LV. Introduction to plant disease Epidemiology. John Wiley and Sons Inc. USA. 1990;532.

18. Otim-Nape GW, Thresh JM, Bua A, Baguma Y, Shaw MW. Temporal spread of cassava mosaic disease in a range of cassava cultivars in different agro-ecological regions of Uganda. Annals of Applied Biology. 1998;133:415-430.

19. Byabakama BA, Adipala E, Ogenga-Latigo MW, Otim-Nape GW. The effect of amount and disposition of inoculum on cassava mosaic virus disease development and tuberous root yield of cassava. African Journal of Plant Protection. 1999;7:45-57.

20. Legg JP, Ogwal S. Changes in the incidence of African cassava mosaic *Geminivirus* and the abundance of its whitefly vector along south–north transects in Uganda. Journal of Applied Entomology. 1998;122:169-178.

21. Byabakama BA. Host resistance and epidemiology of African cassava mosaic virus disease in different agro-ecologies of Uganda. MSc thesis, Makerere University, Kampala; 1996.

22. Adriko J. Evaluation of improved and local cassava varieties in Uganda for resistance/ tolerance to cassava mosaic virus disease (CMD). MSc thesis, Makerere University, Kampala; 2005.

23. Hahn SK, Terey ER, Leuschner K. Breeding cassava for resistance to cassava mosaic disease. Euphytica. 1980;29:673-683.

24. Thresh M, Otim-Nape GW, Fargette D. The components of deployment of resistance to cassava mosaic virus disease. Integrated Pest Management Reviews. 1998;3:209-224.

25. Fishpool LDC, Van Helden M, Van Halder I, Fauquet C, Fargette D. Monitoring *Bemisia tabaci* in cassava: Field counts and trap catches. In *Proceedings of* International Seminar on African Cassava Mosaic Disease and its Control, 4-8 May 1987 (Fauquet, C and Fargette, D. (Eds.). Yamoussoukro, Cote d'Ivoire. CTA, Wagenigen, the Netherlands. 1987;64-76.

26. Fauquet C, Fargatte D, Thouvenel JC. Some aspects of the epidemiology of African cassava mosaic virus in Ivory Coast. Tropical Pest Management. 1998;34:92-96.

27. Mware B, Narla R, Amata R, Olubayo F, Songa J, Kyamanywa S, Ateka EM. Efficiency of cassava brown streak virus transmission by two whitefly species in coastal Kenya. Journal of General and Molecular Virology. 2009;1(4):40-45.

28. Hillocks RJ. Cassava brown streak virus disease: Summary of present knowledge on distribution, spread, effect on yield and methods of control. In: Cassava Brown Streak Virus Disease: Past, Present and the Future. Proceedings of an International workshop, Mombasa, Kenya, 27-30 October 2002. Legg J.P and Hillocks R.J (Eds) 2003. Natural resources International Ltd, Aylesford, UK; 2003.

29. Legg JP, Pheneas N. New spread of Cassava Brown Streak Virus Disease and its implications for the movement of Germplasm in the East and Central African Region. Crop Crisis Control Project; 2007.

30. Patil BL, Legg JP, Kanju E and Fauquet CM. Cassava brown streak disease: A threat to food security in Africa. Journal of General Virology. DOI:10.1099/ jgv.0.000014.

31. Otim-Nape GW, Bua A, Thresh JM, Baguma Y, Ogwal S, Ssemakula GN, Acola G, Byabakama BA, Colvin J, Cooter RJ, Martin A. The current pandemic of cassava mosaic virus disease in East Africa and its control. NARP/NRI/DFID publication, Natural Resources Institute, Chatham, UK. 2000;100.

32. Jeremiah S. The role of whitefly (*Bemisia tabaci*) in the spread and transmission of cassava brown streak disease. PhD thesis. University of Dar es Salaam, Tanzania. 2014;229.

# 10

# Farmers' Perceptions of the Effectiveness of the Cocoa Disease and Pest Control Programme (CODAPEC) in Ghana and Its Effects on Poverty Reduction

**Emmanuel Kumi[1,2*] and Andrew J. Daymond[2]**

[1]*Centre for Development Studies, Department of Social and Policy Sciences, University of Bath, Claverton Down, Bath, BA2 7A, United Kingdom.*
[2]*Policy and Development, School of Agriculture, University of Reading, Whiteknights, Reading, RG6 6AR, United Kingdom.*

***Authors' contributions***

*This work was carried out in collaboration between both authors. Author EK was involved in the study design, data collection and analysis. He managed the literature review as well as the writing up of the introduction and the first draft of the results of the manuscript. Author AJD was involved in the analysis, writing up of the discussion and the proof reading of the manuscript. Both authors read and approved the final manuscript.*

Editor(s):
(1) Juan Yan, Sichuan Agricultural University, China.
Reviewers:
(1) Tuneera Bhadauria, Department of Zoology, Feroz Gandhi College, Kanpur University, Uttar Pradesh, India.
(2) Kanmogne Abraham, Department of Industrial and Mechanical Engineering , National Advanced School of Engineering (University of Yaoundé I), Cameroon.

**ABSTRACT**

The study examined the contribution of the Cocoa Disease and Pest Control Programme (CODAPEC), which is a cocoa production-enhancing government policy, to reducing poverty and raising the living standards of cocoa farmers in Ghana. One hundred and fifty (150) cocoa farmers were randomly selected from five communities in the Bibiani-Anhwiaso-Bekwai district of the Western Region of Ghana and interviewed using structured questionnaires. Just over half of the farmers (53%) perceived the CODAPEC programme as being effective in controlling pests and diseases, whilst 56.6% felt that their yields and hence livelihoods had improved. In some cases pesticides or fungicides were applied later in the season than recommended and this had a

*Corresponding author: E-mail: e.kumi@bath.ac.uk, kumiandy3@gmail.com;*

detrimental effect on yields. To determine the level of poverty amongst farmers, annual household consumption expenditure was used as a proxy indicator. The study found that 4.7% of cocoa farmers were extremely poor having a total annual household consumption expenditure of less than GH¢ 623.10 ($310.00) while 8.0% were poor with less than GH¢ 801.62 ($398.81). An amount of money ranging from GH¢ 20.00 ($9.95) to GH¢ 89.04 ($44.29) per annum was needed to lift the 4.7% of cocoa farmers out of extreme poverty, which could be achieved through modest increases in productivity. The study highlighted how agricultural intervention programmes, such as CODAPEC, have the potential to contribute to improved farmer livelihoods.

*Keywords: Cocoa disease and pest control (CODAPEC); poverty reduction; standard of living; mirids; black pod disease; Ghana.*

## 1. INTRODUCTION

Ghana is the second largest producer and exporter of cocoa beans, after Côte d'Ivoire. In 2012, cocoa accounted for about 30% of Ghana's total export earnings, 19% of agricultural Gross Domestic Product (GDP) and 3.0% of national GDP [1,2]. For Ghanaian cocoa farmers, the contribution of cocoa to annual household income is estimated between 70% to 100% and employs about 3.2 million workers representing 60% of the national agricultural labour force [3]. Smallholder farmers contribute to about 90% of global cocoa production and typically operate within a farm size of 1 to 5 hectares [4]. Similarly, in Ghana, smallholder farmers dominate cocoa production which therefore makes the crop an instrumental vehicle for employment creation and poverty reduction.

From 2003 onwards, the impressive growth performance and poverty reduction recorded in the Ghanaian economy is mainly attributed to the agricultural sector, which is largely driven by cocoa production. Between 2003 and 2007, economic growth rates in Ghana increased from 5.2% to about 6.3% which has resulted in an increase in average income from $1,430 in 2008 to US$2,500 in 2010 [5,6]. The growth in the agricultural sector has been underpinned by the sturdy output performance of the cocoa sector from 0.5% to 16.4% year-on-year growth in output between 2003 and 2012 [7]. However, Ghana's cocoa sector operates at lower yield productivity compared to their counterparts in some countries like Côte d' Ivoire, Indonesia and Malaysia [8]. Research has shown that cocoa farmers in Ghana have the potential to produce an estimated dry bean yield of 1000 kg ha$^{-1}$ or more [9] but currently the national average yield is estimated at 400 kg ha$^{-1}$. The relatively low yield in Ghana is attributed to factors such as high prevalence of pests and diseases, poor agronomic practices, decline in soil fertility and the use of low yielding varieties [10].

Among the above factors, the impact of pests and diseases is one of the greatest challenges to farmers. In Ghana, common cocoa diseases include *Phytophthora* black pod caused by the species *Phytophthora palmivora* and *Phytophthora megakarya* and Cocoa Swollen Shoot Virus Disease (CSSVD) while that of pests include insects mostly of the bug or miridiae family such as *Distantiella theobroma* and *Sahlbergella singularis*. Such diseases and pests can have devastating effects on the economies of cocoa production by reducing yields [11]. Although difficult to quantify, Acebo-Guerrero et al. [12] have argued that *Phytophthora megakarya* and mirids could cause an estimated 70%-90% annual crop loss if control measures are not taken and consequently significant economic loss to farmers. For example, between 2008 and 2010, an average estimated value of more than US$300 million of annual crop loss in Ghana was attributed to black pod disease while loss due to mirids infection was estimated at US$172 million [13]. The loss in productivity translates into low income which implicitly affects the standards of living of farming households. This invariably creates apathy on the part of farmers in making productive investments such as the use of fungicides, insecticides and fertilisers on their farms. Consequently, the long term growth and sustainability of the cocoa sector is threatened [14].

In addressing the challenges of the cocoa sector, a number of policies, programmes and interventions aimed at improving farm level productivity among farmers such as the Cocoa High-Technology Programme (Cocoa Hi-Tech) have been implemented over the years, with the aim of improving the livelihood of farmers [15]. In 2001 the government of Ghana initiated the Cocoa Disease and Pest Control Programme

(CODAPEC), a national cocoa spraying programme with the objective of facilitating increased production among farmers through the control of mirids and black pod disease. Over the years, yield improvements in the cocoa sector have been linked at least in part to CODAPEC. For example, the level of national cocoa output increased from 632,000 to 1,025,000 metric tonnes for 2009/2010 and 2010/2011 growing seasons respectively. This was accompanied by an increase in the farm gate producer price from GH¢ 2,400 to GH¢ 3,200 per metric tonnes of cocoa beans [16]. It has been argued that the increased production levels and the price incentive has led to an increase in farmers income and has therefore resulted in the reduction of poverty among cocoa-farming households [17].

In Ghana, poverty abounds especially among rural farming households, even though, poverty according to the Ghana Living Standard Survey 5 is said to have reduced at an unprecedented rate from 51.7% to 28.5% between 1991 and 2005 [18]. However, since the inception of CODAPEC in 2001, relatively little is known about the poverty levels and standards of living among cocoa farmers. This study was conducted in the Bibiani- Anhwiaso Bekwai District (hereafter, BABD) in the Western Region of Ghana. Poverty is a common phenomenon that is often experienced in the district as 35.1% of farmers are classified as extremely poor [19]. Similarly, Boon et al. [19] found that about 58% of the population of BABD lives under the national poverty line. This raises key research questions such as: Has the living standard of cocoa farmers improved since the implementation of CODAPEC? What percentage of farmers can be classified as living below the upper and lower poverty lines? What is the perception of cocoa farmers of their living conditions? Despite the previous studies on CODAPEC and cocoa production in Ghana, there is relatively little empirical research that focuses directly on assessing the effects of the programme on poverty reduction and standards of living among cocoa farming households in the context of Ghana.

This paper therefore attempts to analyse the poverty levels and standards of living of cocoa farmers in the BABD of the Western Region of Ghana by using farmers' household expenditure in 2012 as a proxy indicator in comparison with the national poverty lines set by the Ghana Statistical Service [20]. The paper also explores farmers' perceptions of the CODAPEC programme and identifies policy gaps in the implementation of CODAPEC. We take cognisance of the fact that in taking an instrumental view of asking project beneficiaries or farmers directly about attribution or their perception of the programme, there is a possibility of confirmation bias [21] or what Copestake [22] calls 'project bias' where "someone consciously or otherwise conceals or distorts what they think they know about an activity in the hope that doing so will reinforce the case for keeping it going". Despite this potential, *ex post* consultation or econometric evaluations cannot be used as a substitute for cocoa farmers in whose name the programme was implemented. As Copestake [22] argues, it is ethically correct to involve at least some direct beneficiaries in project evaluation even if it presents methodological challenges.

Up to date, research on cocoa farmers and CODAPEC in Ghana is largely skewed towards quantitative approach which focuses mainly on an axiomatic view of the project while others have also focused on factors that influence the adoption of the programme by farmers [23]. In this paper, we argue that such a view is insufficient, thus creating a gap in knowledge about farmers subjective assessment of the programme. There is therefore little empirical research on the subjective evaluation of the programme with respect to farmers' perception and its effects on their households and standards of living in general. This research seeks to fill this scholarly gap. The novelty of the paper lies in its potential contribution in deepening our understanding about the effects of government agricultural policy initiatives on the livelihood of cocoa farming households of which little is known.

## 2. MATERIALS AND METHODS

### 2.1 Study Area

The study was conducted in the BABD located at the North-western part of the Western Region of Ghana. The district, which is located between latitude 6° N, 3° N and longitude 2° W, 3° W (Fig. 1) is an important cocoa producing area in Ghana and covers an estimated land area of 873 km². An estimated 62% representing 39,829 hectares out of the 54,240 hectares of the available total arable land is under cultivation for both cash and food crops such as cocoa, coffee, plantain and cassava [24,25]. Topographically, the land rises from about 350m to 660m above sea level [26]. The district is located in the wet-semi equatorial rainforest zone which is marked by a bimodal rainfall pattern between March-August and September-October. Average annual rainfall is between 1200mm and 1500mm with the peak periods between June and October [27]. The dry season is between November and January. BABD has a uniform average temperature of around 26°C throughout the year with relative high humidity, daily averages being between 75% and 95% [24]. The favourable climatic conditions combined with the high fertility of forest ochrosols soil supports cocoa production and makes it the most important cash crop cultivated by farmers [20,27]. Furthermore, food crops like plantain, cassava, rice and black pepper are also cultivated on an average farm size of 1.5 hectares [25].

BABD is basically agrarian with an estimated 61% of the active labour force engaging in agricultural activities such as crop and mixed farming in addition to animal husbandry. Mining activities in gold and bauxite in Bibiani, Chirano and Awaso respectively dominates the industry sector. BABDA's population according to the 2010 Housing and Population Census of Ghana was estimated at 123,272 with 49.4% male and 50.6% female [24].

## 2.2 Sampling Procedure, Data Collection and Analysis

Purposive sampling was used in selecting the BABD for the study. Cluster sampling was also used in dividing the district into three zones: Bibiani Zone, Anhwiaso Zone and Bekwai Zone because of the expansive nature of the district. Simple random sampling was further employed in selecting five communities; Kwamekrom (6° N, 25° N, 2° W, 18° W), Dominibo No. 2 (6° N, 21° N, 2° W, 16° W), Tanoso (6° N, 20° N, 2° W, 18° W), Ntakam (6° N, 16° N, 2° W, 19° W) and Humjibre (6° N, 08° N, 2° W, 16° W) (Fig. 1). We then employed the random sampling technique in selecting 150 cocoa farming households, 30 from each community. The same sample size was used for the three zones because of similar population characteristics among cocoa farmers. Data for the study were obtained through the administration of structured questionnaires which were made up of both closed and open-ended questions. Open-ended questions were used to capture qualitative data representing the respondent's own views about their household

expenditure and standards of living while the closed-ended questions elicited information for the quantitative analyses [28]. The questionnaires were pre-tested in two communities; Domino No. 1 (6° N, 22° N, 2° W, 17° W) and Bibiani Old Town (6° N, 27° N, 2° W, 19° W) (Fig. 1). The pre-test survey was used to test the feasibility of the questionnaire [28]. Corrections were made to the questionnaire after the pre-test exercise in order to ensure that there was no ambiguity in the questions asked. A team of enumerators pre-tested and administered the questionnaires in the local dialect of respondents. Data collected included those on socio-economic characteristics and demographics, detailed household income and expenditure, the perception of farmers about their living standards and the effectiveness of CODAPEC in improving yield and income.

Descriptive statistics in the form of frequencies, percentages as well as pictograms such as pie charts and bar charts were used to present data whilst associations between socio-economic characteristics and yield were analysed by means of chi-square using the Statistical Product and Service Solutions (SPSS) package, version 20.0. Annual household expenditure in Ghana Cedis (GH¢) was converted into United States Dollars (US$) based on the prevailing market exchange rate (US$ 1= GH¢ 2.01) in June, 2013. The results were compared to the dollar equivalent of the upper and lower poverty lines set by the Ghana Statistical Service [18] and also from poverty lines calculated from the minimum wage index. The average interbank exchange rate for June, 2006 was at $1= GH¢ 0.92.

## 3. RESULTS

### 3.1 Socio-economic Characteristics of Respondents

#### 3.1.1 Gender profile, marital status and age of respondents

The results of the descriptive statistics of the socio-economic characteristics of 150 cocoa farmers are presented in Table 1. The results indicated a high ratio of male (62%) to female (38%) farmers. About 82.7% of respondents were married or had married before but are currently divorced, living in consensual union or widowed. The average age of respondents was about 40 years with the 31-40 years age bracket being the modal age class. The age of farmers ranged from 18 to 70 years. The results demonstrated a fair distribution of ages across the population with majority of farmers (92.7%) being in their economically active age (18-64 years).

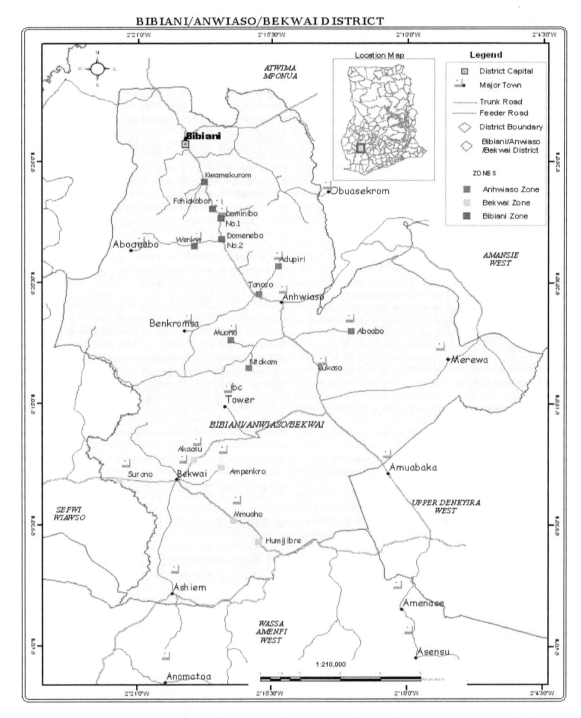

**Fig. 1. Bibiani-Anhwiaso-Bekwai district showing location and selected communities for the study**

**Table 1. Descriptive statistics of the socio-economic characteristics of respondents (n=150)**

| Socio economic variable | Frequency | Percentage |
|---|---|---|
| **1. Gender profile of respondents** | | |
| Male | 93 | 62.0 |
| Female | 57 | 38.0 |
| **Total** | **150** | **100.0** |
| **2. Marital status of respondents** | | |
| Married | 78 | 52.00 |
| Single | 26 | 17.3 |
| Divorced | 16 | 10.7 |
| Widowed | 27 | 18.0 |
| Consensual Union | 3 | 2.0 |
| **Total** | **150** | **100.0** |
| **3. Experience in cocoa cultivation** | | |
| Less than 5 years | 25 | 16.7 |
| 5-10 years | 33 | 22.0 |
| 11-20 years | 47 | 33.1 |
| 21-30 years | 27 | 18.0 |
| Above 30+ | 18 | 12.0 |
| **Total** | **150** | **100.0** |
| **4. Educational level of respondents** | | |
| Basic (Primary and Middle School) | 35 | 23.3 |
| Secondary (Senior High School) | 29 | 19.3 |
| Tertiary | 11 | 7.4 |
| No education | 44 | 29.3 |
| Non formal education | 31 | 20.7 |
| **Total** | **150** | **100.0** |
| **5. Household size** | | |
| 1-5 member(s) | 82 | 54.7 |
| 6-10 members | 58 | 38.7 |
| 11-15 members | 8 | 5.3 |
| 16-20 members | 2 | 1.3 |
| **Total** | **150** | **100.0** |
| **6. Number of household members working on farm** | | |
| 1-2 person (s) | 100 | 66.7 |
| 3-4 persons | 39 | 26.0 |
| 5-6 | 10 | 6.6 |
| Above 6 persons | 1 | 0.7 |
| **Total** | **150** | **100.0** |

*Source: Field survey, 2013*

### 3.1.2 Cocoa farmer's level of education and experience

The adult illiteracy rate (percentage of persons aged 15 years and over who cannot read and write) was found to be 29.3% although 23.3% were educated to the basic school level while a few had attained tertiary education. The average working experience was about 15 years while the years of experience in cocoa farming ranged from 5 to 30 years (Table 1). There was a highly significant relationship between respondent's experience in cocoa cultivation and their yield ha$^{-1}$ ($X^2$ = 70.50, $P$=<0.01). Farmers with more years of experience in growing cocoa had higher yield ha$^{-1}$ compared to farmers with less experience.

### 3.1.3 Farm size and cocoa output

Fig. 2 presents the distribution of cocoa farm sizes as reported by farmers. About 68.7% of surveyed farmers claimed to have farm size between 1.0 and 4.0 hectares while a small proportion (2%) had above 10 hectares. Results indicate that smallholder farmers dominate cocoa farming in the study area. The average farm was 1.6 hectares, with the range being from 0.40 to 15 hectares.

Results on cocoa output (64 kg/Bag) and kg ha$^{-1}$ produced by farmers are presented in Table 2. The results show that farmers had an average yield of 574 kg ha$^{-1}$, the range being from 300 kg ha$^{-1}$ to 685 kg ha$^{-1}$.

**Table 2. Descriptive statistics of the output of cocoa (Bags) and (Kg ha$^{-1}$) of the sampled cocoa farmers (n=150)**

| Variable | Frequency | Percentage |
|---|---|---|
| **Output of cocoa (64 kg/ bag of dry beans)** | | |
| < 10 | 26 | 17.3 |
| 10.5 – 20 | 51 | 34 |
| 20.5 – 30 | 31 | 20.7 |
| 30.5 – 40 | 22 | 14.7 |
| 40.5 – 50 | 4 | 2.6 |
| Above 50+ | 16 | 10.7 |
| **Total** | **150** | **100.0** |
| **Output of cocoa (Kg ha$^{-1}$)** | | |
| Less than 300 | 12 | 8.0 |
| 300-400 | 15 | 10.0 |
| 401-500 | 34 | 22.7 |
| 501-600 | 61 | 40.7 |
| Above 600 | 28 | 18.6 |
| **Total** | **150** | **100.0** |

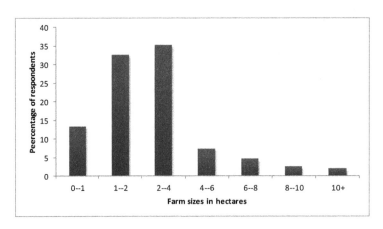

**Fig. 2. Distribution of farm size hectares (n=150)**

### 3.1.4 Variations in the first month of spraying

Fig. 3 presents the responses of farmers on the timing of the first pesticide/fungicide spray. A majority of respondents (94.7%) had their farms sprayed within the months of July, August, September and November, which is the recommended spraying period. However, a smaller proportion (5.3%) reported that their farms were sprayed beyond November. The results of the Chi-square tests statistic ($X^2$ =228.68; $P$=0.04) on the association between the period of spray and yield ha$^{-1}$ shows a statistically significant relationship at 5% level of significance. The highest yields were recorded for farms that were sprayed in the month of July while farms had lower yields when sprayed in November.

### 3.1.5 Farmers' perception on the effectiveness of the spraying process and economically important pests and diseases

A large proportion (53%) of farmers claimed that the spraying process was effective in controlling the incidence of pests and diseases. However, a much smaller proportion (14%) claimed spraying under COPAPEC was ineffective as the programme is faced with numerous institutional constraints.

Fig. 4 shows the pests and diseases reported by farmers in terms of economic importance. A large proportion of farmers identified mirids and black pod (45% and 23% respectively) as the most economically important pest and disease. Cocoa swollen shoot virus disease was cited by 17% of respondents as the most important disease, whilst 10% cited mistletoe growth in the cocoa canopy.

### 3.1.6 Inefficiencies and challenges facing CODAPEC

Fig. 5 illustrates the key inefficiencies and challenges of CODAPEC identified by farmers and spraying gangs. The untimely supply of insecticides and fungicides was cited by both by farmers (26%) and spraying gangs (50%) as the major challenge. The perception of sprayers on the programme was also sought since they are the workers on the ground. This helped in providing a deeper understanding of the challenges that confronts them in undertaking their spraying exercise. Their perception on the challenges facing CODAPEC is presented in Fig. 5.

## 3.2 Effect of CODAPEC on Crop Yield, Household Income and Standards of Living

### 3.2.1 Farmers perception on the effect of CODAPEC on crop yield

The perception of farmers on the effect of CODAPEC on crop yield was assessed by asking the respondents to compare their yield before and after the implementation of CODAPEC. The majority (56.7%) of respondents claimed the spraying exercise was effective in increasing the yields of cocoa. About 17.3% of respondents claimed there have been no significant variations in their yields, whilst 20% claimed their yields had decreased. A statistically significant relationship was found between month of first spray and farmer's perception of an increase in cocoa yields since the inception of CODAPEC ($X^2$ = 59.59; $P$=<0.01). Farmers who reported late spraying tended not to see a yield advantage ($X^2$= 23.6; $P$=0.75).

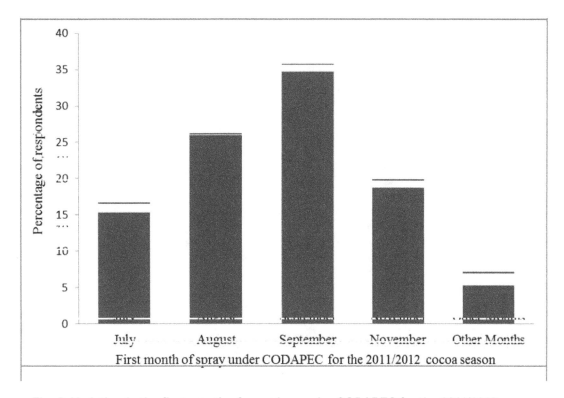

**Fig. 3. Variation in the first month of spraying under CODAPEC for the 2011/2012 cocoa growing season (n=150)**

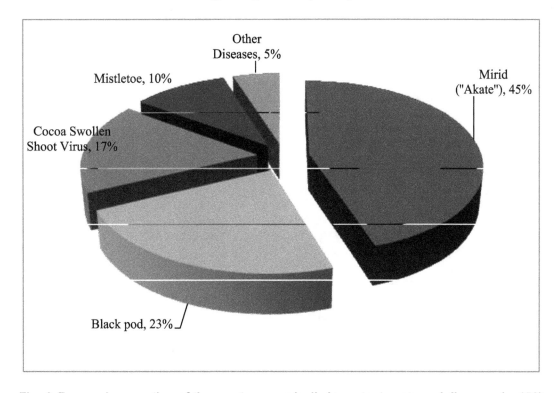

**Fig. 4. Farmers' perception of the most economically important pests and diseases (n=150)**

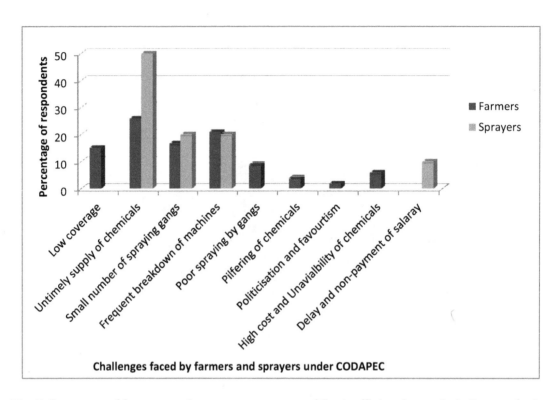

**Fig. 5. Summary of farmers and sprayers response of the inefficiencies and challenges facing CODAPEC (n=160)**

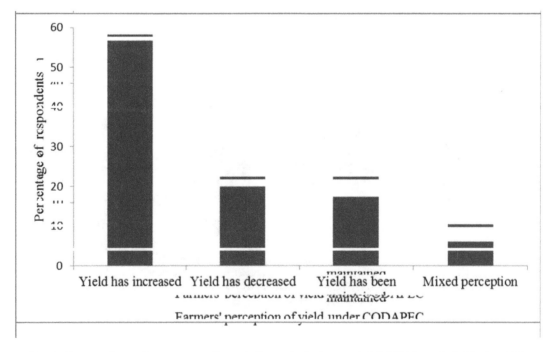

**Fig. 6. Perception of farmers on the relationship between CODAPEC and yields (n=150)**

### 3.2.2 Sources and proportion of household income from cocoa farming

The mean annual household income from cocoa was GH¢5,073.75 ($2,525.25), the range being from GH¢ 615.00 to 16,400.00. Cocoa farming was the main occupation of respondents and accounting for 75.3% of total household income on average. Sales from cocoa beans were cited as the highest income source for farmers in addition to food crops (Table 3). Households were highly reliant on the income from cocoa.

### 3.2.3 Household consumption and expenditure

The household expenditure excluding on-farm expenditure level is presented in Table 4. Food expenditure accounted for a large proportion (45.2%) of the household total expenditure while a small proportion (6.1%) was spent on education and health.

### 3.2.4 Comparative analysis of average and total household consumption expenditure to Ghana living standards survey (GLSS 5)

Poverty levels of respondents were determined by using household consumption expenditure as a proxy indicator. The mean household expenditure from the study was compared to that of the Ghana Living Standard Survey (GLSS) 5 mean household expenditure GH¢ 1,918.00 ($2,084.78) in 2006 (GSS, 2008). Results from Table 5 indicates that respondents could spend more money GH¢ 3,383.00 ($1,683.08) and could be suggested that monetary wise, they are better off in 2013 than in 2006 in terms of Ghana Cedis (GH¢). The average expenditure is about 1.7 and 2.0 times more than the national and rural forest average household expenditure.

Table 6 presents the percentage of respondents living below the national poverty line computed from the upper and lower poverty lines which were GH¢ 370.90 and 288.50 per adult per year respectively in 2006 [29]. When these figures were inflated to 2013 levels by adjusting for the change in exchange rates, they were GH¢ 623.10 and GH¢ 801.62 respectively. Based on the annual household consumption expenditure, 4.7% and 8.0% of farmers are also classified as extremely poor and poor respectively in accordance with national poverty lines. When comparing the household consumption expenditure with the Minimum Wage Index used in Ghana, 14% of respondents were classified as

living in poverty. An amount of money ranging from GH¢ 20.00 to GH¢ 89.04 per annum is needed to lift the 4.7% of respondents out of extreme poverty to the poor line when using the poverty gap.

### 3.3 Farmers Perception about Their Poverty Level and Standards of Living

Fig. 7 illustrates the perception of farmers about their standard of living based on the multidimensionality of poverty approach (defining poverty in terms of non-income dimensions of human well-being and people's own lived experiences). Respondents were asked about their poverty levels and living conditions on a scale of four, from very good standards to very poor standards of living.

The results illustrated in Fig. 7 suggest a higher level of perceived poverty than those presented in Table 6. About 11% and 6% of farmers considered themselves to be poor and extremely poor respectively.

### Table 3. Proportion of household income from cocoa farming (n=150)

| Sources of household income | Frequency | Percentage |
|---|---|---|
| Food stuffs (Plantain, Cassava, Yam, Cocoyam) | 23 | 15.30 |
| Cocoa Beans | 113 | 75.30 |
| Vegetables (Pepper, Tomatoes etc.) | 14 | 9.30 |
| Total | 150 | 100.0 |

## 4. DISCUSSION

### 4.1 Socio-economic Characteristics of Respondents

The current study has demonstrated the dominance of an economically active farmer population in the production of cocoa in the BABD district of Ghana. Results from the present case study area are consistent with the findings of Danso-Abbeam et al. [30] who reported that about 61% of cocoa farmers in the BABD were aged between 20 and 50 years. The results are however in contrast with studies from other parts of the country that indicates an aging farming population [23,31,32]. As cocoa farmers are ageing, there is the need for a replacement by younger farmers to ensure the sustainability of the cocoa sector. Programmes that are geared towards improving farmer yields such as CODAPEC have a potential role in encouraging future generations to continue cocoa farming.

**Table 4. Household annual expenditure (GH¢) for the 2012/2013 cocoa growing season (n=150)**

| Expenditure Item | Number of farmers | Minimum | Maximum | Mean | Median | Std. deviation |
|---|---|---|---|---|---|---|
| Root & tuber crops | 150 | 50.00 | 5,000.00 | 1,188.41 | 864.00 | 1,176.34 |
| Bread and cereals | 150 | 15.00 | 900.00 | 150.50 | 100.00 | 175.29 |
| Meat and fish | 150 | 30.00 | 980.00 | 160.23 | 100.00 | 170.02 |
| Oil, fats, vegetables | 150 | 5.00 | 120.00 | 30.53 | 25.00 | 22.43 |
| Clothing & footwear | 150 | 10.00 | 1,500.00 | 257.94 | 150.00 | 264.95 |
| Charcoal & gas | 150 | 10.00 | 300.00 | 49.34 | 37.50 | 41.64 |
| Water & electricity | 150 | 20.00 | 750.00 | 168.31 | 100.00 | 164.02 |
| Rental& housing | 150 | 10.00 | 700.00 | 173.36 | 137.50 | 122.34 |
| Toiletries | 150 | 10.00 | 500.00 | 141.76 | 100.00 | 110.91 |
| Funerals | 150 | 10.00 | 2,000.00 | 248.06 | 150.00 | 287.30 |
| Transport & Comm. | 150 | 10.00 | 1,700.00 | 194.63 | 110.00 | 259.37 |
| Church | 150 | 10.00 | 1,000.00 | 168.48 | 120.00 | 164.24 |
| Health | 150 | 10.00 | 540.00 | 91.52 | 70.00 | 82.76 |
| Education | 150 | 10.00 | 600.00 | 116.50 | 100.00 | 98.68 |
| Miscellaneous | 150 | 10.00 | 4,500.00 | 243.43 | 120.00 | 427.42 |
| **Total** | **150** | **220** | **24,090.00** | **3,383.00** | **2,284.00** | **3,567.71** |

**Table 5. Comparison of mean total expenditure between GLSS 5 (2006) and field data (2013)**

| Household expenditure | GLSS 5 (June, 2006)* | Field survey (June, 2013) ** |
|---|---|---|
| Mean household expenditure in Ghana | GH¢ 1,918.00 ($2,084.78) | - |
| Mean expenditure for rural forest zone | GH¢ 1,629.00 ($1,770.65) | - |
| Mean household expenditure (field survey) | - | GH¢ 3,383.00 ($1,683.08) |

*Source: Authors' calculation using data from the GLSS 5 and Field Survey.*
*\* The average interbank exchange rate for June, 2006 was at GH¢1=$ 0.92.*
*\*\* The average interbank exchange rate for June, 2013 was at GH¢1=$ 2.01*

**Table 6. Percentage of respondents living below the national poverty line in 2013 based on computed and updated from the national living standard survey 2006 (GLSS 5)**

| GLSS 5 poverty lines (2006)[+]per equivalent adult (GH¢) GH¢ 1 = $0.92 | Computed GSS poverty lines in 2013 [++] (GH¢) | Percentage of respondents living below poverty lines (field survey, 2013)[+++] | Minimum wage index as at June 2013 GH¢ 5.24 /day * 264 working days per year (GH¢) [++++] | Percentage of population below minimum wage in 2013 |
|---|---|---|---|---|
| Extremely poor  288.50 | 630.29 (288.50/0.92) *2.01 | 4.7% | GH¢ 1, 383.36 | 14% |
| Poor  370.90 | 810.33 (370.90/0.92) *2.01 | 8.0% | | |

*Source: Authors' calculation using data from the GLSS 5 and Field Survey, 2013.*
*+ Two poverty lines used in the GLSS 5.++ Inflated value of GLSS 5 Upper and Lower poverty lines at the exchange rate as at June, 2013 ($1 = GH¢ 2.01).*
*+++ Percentage of respondents living below computed the national poverty lines based on farmers income.*
*++++Workers in Ghana work for 5 days a week so it is assumed that there are 22 working days in a month for each of the 12 months in a year (264 working days)*

The current study has demonstrated that a large proportion of cocoa farmers are literate with a majority attaining basic education and thus is consistent with the observations of Baah et al. [10] and Aneani et al. [31] who found an increasing literate population among cocoa farmers.Despite this improvement, the proportion of farmers with qualifications above secondary level was found to be low. The results presented here are consistent with the claim by Asamoah et al. [33] that cocoa farmers with qualifications above secondary level is less than 5% in Ghana. The situation may have implications on the efficient and effective use of innovations including pests and diseases control measures. It was notable that more experienced farmers achieved higher yields. Cocoa cultivation in Ghana provides an important income source to farmers thereby helping to improve the standard of living and reducing poverty among farming households. However, productivity among cocoa farms in Ghana is relatively low [34] and thus the potential income of cocoa farmers is not always realised. The average cocoa yield of 574 kg ha$^{-1}$ presented in this study is higher than the 378.81 kg ha$^{-1}$ reported by Danso-Abbeam et al. [30] in the same study area and the national average of 400 kg ha$^{-1}$ [10].

## 4.2 Farmer's Perception on CODAPEC

Farmers perceived mirids to be the most economically important group of pests that threaten cocoa cultivation in the study area followed by black pod disease and Cocoa Swollen Shoot Virus Disease (CSSVD). Mirids (Sahlbergella and Distantiella) are reported to cause considerable damage to cocoa production resulting in loss of yield of about 25% whereas losses from blackpod can sometimes reach 70-90% in Ghana if appropriate measures are not taken [10]. In the Western Region of Ghana, CSSVD can cause complete crop loss when infection is severe [35]. Mistletoe infestation is also gaining prominence as an impediment to cocoa production as it causes reduced yields and deterioration of farms through canopy damage and the promotion of mirid pockets [36].

Timing of pesticide application is critical to maximise its effectiveness in controlling mirids. The mirid population in West Africa, starts to build-up in July and reaches its peak between August and September while black pod occurrence increases from June with peaks in August and October. Consequently, it is recommended that cocoa farms in Ghana are sprayed between July and September. As indicated in this study, the majority of farmers had their farms sprayed between July and September but a significant proportion received their first spraying under CODAPEC in September when the population of *Sahlbergella singularis* would have been at its peak and therefore already caused damage to the crop. Surprisingly, some farmers had their farms sprayed in November. In these cases pod loss due to mirids would have already peaked before farms were sprayed. The results from the present study mirror the situation in the Ahafo-Ano and Upper Denkyira districts of Ghana where Abankwah et al. [11] and Anang et al. [27] found that 59% and 30% of farmers respectively had their farms sprayed in September and beyond due to delay in the supply of chemicals.

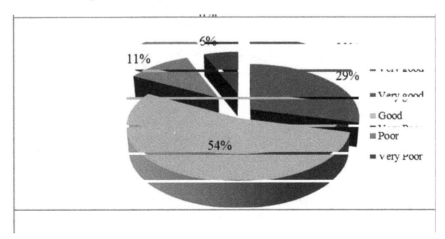

**Fig. 7. Farmers perception about their poverty level and standards of living (n=150)**

The effective implementation of CODAPEC in the control of pests and diseases was shown to be hindered by several challenges among which include the untimely supply of chemicals, as also shown by Aneani et al. [31] and Helmsing and Vellema [37]. Additionally, farmers cited the inadequate spraying personnel as a major challenge that confronts all the community in the study areas. Farmers maintained that this situation resulted in delays to spraying; furthermore, 10% of farms were not covered at all. This suggests that greater attention needs to be focused on employing more spraying personnel for the effective implementation of the programme in the study area. Moreover, pilfering of chemicals and poor spraying by spraying gangs was also a concern to farmers. The findings here also illustrate administrative lapses in the monitoring and evaluation aspects of the programme in the district.

This assertion by farmers confirms that of desk and extension officers who acknowledged that financial, logistical and infrastructural constraint limit their ability in carrying out effective monitoring. Nevertheless, sprayers are monitored through field visits, engagement with farmers and the submission of reports by gang leaders. Similar results have been found by Teal et al. [38] in the Western region. The results presented here highlight the importance of strengthening the monitoring and evaluation component of CODAPEC at the district and community level.

### 4.3 Effect of CODAPEC on Crop Yield, Household Income and Standards of Living

A large proportion (56.7%) of farmers perceived that CODAPEC has led to an increase in their productivity (yield) over the years as a result of the reduction in the incidence of pests and diseases. Thus, a majority of farmers had a positive impression about the programme. However, a significant proportion of framers did not perceive a yield advantage (20% of farmers thought that yields had actually decreased). It would appear that in these cases late application of pesticide resulted in little or no yield advantage. It is important to state that spraying is not the only factor that improves farm productivity; other factors such as husbandry practices also influence productivity as well as year-to-year changes in climate. The findings here are broadly consistent with observations by Opoku et al. [39] that since the inception of

CODAPEC, the majority of farmers in Ghana have testified that cocoa yields have increased on their farms. Adu-Ampomah et al. [35] found that yields of cocoa increased from 266 kg ha$^{-1}$ to 434 kg ha$^{-1}$ between 2001 and 2003 and attributed the success to CODAPEC. Farmers report and the literature agrees that cocoa production has continued to be the primary source of income for most cocoa farming household in Ghana [38]. A large proportion of households (75.3%) were mostly reliant on the income from the sale of cocoa beans for their upkeep. Daniel et al. [40] stated that cocoa income often serves as the only readily available income source for farming households especially when household needs such as food, education and social contributions such as those on funerals and church activities are to be met.

However, over dependence on cocoa income during the off-season periods could have negative consequences on household income levels in the event that yields and/or the cocoa price fall. As a risk averting strategy, farmers have diversified their income sources through the growing of other crops (e.g. palm) and non-farm activities such as trading. This indicates the importance of income diversification to household livelihoods especially in the rural areas of Ghana [41]. It is frequently asserted that farmers are expected to be food secured at the household level as they meet a large proportion of their consumption through subsistence production [42].

However, the results indicated that cocoa farmers spend a greater proportion of their income on food expenditure. This raises some concern about the household level food security as an increase in food prices might affect households that purchase the majority of their food from the market pushing more farming households into food shortage, deprivation and poverty [43]. Surprisingly, expenditure on health and education constituted a smaller proportion of household total expenditure. Claims were made by some farmers that the social intervention programmes by the central government such as the National Health Insurance Scheme (NHIS), Free Compulsory Universal Basic Education and the School Feeding programmes have lessened their social household expenditure.

Defining and measuring the concept of poverty is a complex task. While there are several indicators for measuring poverty, they are not without problems. A widely acceptable measure of household welfare (poverty levels and standard of living) has been the consumption expenditure [44]. Household consumption expenditure was used as a proxy indicator of poverty levels because of the difficulties involved in measuring farmers' income. Based on the computed poverty line in 2013 (Table 6), the finding indicates that 4.7% and 8.0% of respondents can be classified as being "extremely poor" (living below the lower poverty line) and "poor" (living below the upper poverty line) respectively. Thus, the percentage of farmers who were classified as 'extremely poor' and 'poor' in relative terms had an annual household consumption expenditure below the national poverty lines of GH¢ 801.62 and GH¢ 623.10 respectively. In monetary terms when using the Minimum Wage Index in determining the levels of poverty, 14% of respondents can be considered to be living in poverty.

From a monetary perspective, poverty in Ghana has been decreasing such that the number of people defined as living in poverty fell by 1.6 million between 1992 and 2006 [45]. The country is said to be on track to achieving the Millennium Development Goal (MDG One) of halving the poverty rate by 2015. This is backed by the growth of the economy which has been aided by an increase in world commodity prices of exports including that of cocoa over recent years [46,47,48]. Economic growth trends in terms of real Gross Domestic Product recorded an increase from 4.7 to 5.9 between 2008 and 2010 [45]. The increased growth is often assumed to have translated into improved standard of living and wellbeing. The reduction in poverty is a result of a number of policies such as Ghana Shared Growth and Development Agenda and the Growth and Poverty Reduction Strategy (GPRS) I and II. Through these policies among many others, the national poverty rate was halved from 51.7% to 28.5% between 1999 and 2005 although with persistent inequality across regions [6,49,50].

When considering specifically rural poverty, the current findings support those of Coulombe and Wodon [51] who found a reduction in rural poverty headcount by 39.2% between 1998/99 and 2005/06 especially in coastal and forest areas. The result from the study confirms the declining poverty trends among cocoa farming households in Ghana. A similar observation has been reported by the Ghana Statistical Service [18] that poverty levels among rural households especially those in forest areas declined from

64% to 28.5% between 1991 and 2005. Similarly, declining poverty levels have been observed among cocoa farmers in the BABD [18]. From the current study, an analysis of the poverty gap (the distance that separates the population from the poverty line) indicates that an amount of money between GH¢ 20.00 ($9.95) to GH¢ 89.04 ($42.29) is needed to lift the 4.7% of farmers to the lower poverty line. Such an uplift in income could be achieved by strengthening yield enhancing programmes such as CODAPEC and Cocoa Hi-Tech so as to increase farm productivity among farmers. It is important to acknowledge that using only income or consumption dimension as a proxy measure of economic well-being and standards of living can be limiting. People experience poverty in different ways, some which cannot be captured in monetary terms, such as access to running water and to electricity [52].

Multidimensional poverty goes beyond the economic well-being measurement and defines poverty in terms of human well-being and people's own lived experiences. The study therefore sought to find out farmers perception about their poverty levels and living standards since they are the people who can really define their own poverty [53,54]. The non-income dimension measurements revealed that about 17% of farmers considered themselves to be poor. This finding reinforces the point made by Whelan et al. [55] about the empirical mismatch between income poverty and people's perception of their own poverty levels and living standards. A similar finding has been reported by Mitra et al. [56] who found in South Africa that living standards are associated with multidimensional poverty rather than income. Furthermore, the study supports the premise that poverty is multidimensional and has many correlates that include both income and non-income aspects of well-being such as assets owned and services received by households [57].

We acknowledge that it might be difficult to establish causality between CODAPEC, poverty reduction and improved living standards among cocoa farmers because of other external factors (agronomic and socio-economic). Nonetheless, since the interest of the study was on the subjective evaluation of farmers, there is evidence to suggest that for a significant proportion of farmers, CODAPEC in addition to other socio-economic and agronomic factors has led to a reduction in poverty through an increase in the farm productivity (yields) of cocoa which reflects the higher incomes earned by farmers. A complement of this finding was seen in the large proportion of respondents who claimed their living standards were good and very good respectively. The higher number of people who claim to have witnessed an improvement in their living conditions could also in part be as a result of the provision of public goods such as scholarship and also bonuses supplied by COCOBOD as a way of attracting farmers into cocoa production.

## 5. CONCLUSION

To conclude, a large proportion of smallholder farmers in Ghana are heavily dependent on the sale of their cocoa beans for their livelihoods. Thus agricultural intervention programmes such as the CODAPEC initiative explored here can partially contribute to reducing poverty and improvements in the living standards cocoa farmers.

## ACKNOWLEDGEMENTS

The authors are grateful to the cocoa farmers who served as study participants. We gratefully acknowledge Mr. John Kwegyire who provided diverse assistant during data collection process. We also want to thank Dr. Alistair Murdoch, University of Reading and the peer reviewers for their constructive feedback.

## COMPETING INTERESTS

Authors have declared that no competing interests exist.

## REFERENCES

1.  Quarmine W, Haagsma R, Sakyi-Dawson O, Asante F, van Huis A, Obeng-Ofori D. Incentives for cocoa bean production in Ghana: Does quality matter? NJAS - Wageningen Journal of Life Sciences. 2012;60–63(0):7-14.

2.  Ghana Statistical Service (GSS). Ghana living standards survey: Report of the Fifth round. Accra: Ghana; 2008. Accessed 8 August 2013.
    Available:http://www.statsghana.gov.gh/docfiles/glss5_report.pdf

3.  Anang BT, Mensah F, Asamoah A. Farmers' Assessment of the Government Spraying Program in Ghana. Journal of Economics and Sustainable Development. 2013;4(7):92-9.

4.  Duguma B, Gockowski J, Bakala J. Smallholder cacao (*Theobroma cacao* Linn.) cultivation in agroforestry systems of West and Central Africa: Challenges and opportunities. Agroforestry systems. 2001;51(3):177-88.

5.  Institute of Statistical, Social and Economic Research (ISSER). The state of the Ghanaian economy in 2007. ISSER, University of Ghana, Ghana; 2008.

6.  Quartey P. Migration in Ghana: A country Profile 2009. International Organization for Migration Geneva Switzerland; 2009.

7.  Onumah JA, Al-Hassan RM, Onumah EE. Productivity and Technical Efficiency of Cocoa Production in Eastern Ghana. Journal of Economics and Sustainable Development. 2013;4(4):106-17.

8.  Jano P, Mainville D. The cacao marketing chain in Ecuador: Analysis of chain constraints to the development of markets for high-quality cacao: IAMA submission 2007 Conference Parma, Italy; 2007.

9.  Aneani F, Anchirinah VM, Asamoah M, Owusu-Ansah F. Baseline socio-economic and farm management survey. A Final Report for the Ghana Cocoa Farmers' Newspaper Project. New Tafo-Akim, Ghana: Cocoa Research Institute of Ghana (CRIG); 2007.

10. Baah F, Anchirinah, V. Amon-Armah, F. Soil fertility management practices of cocoa farmers in the Eastern Region of Ghana. Agriculture and Biology Journal of North America. 2011;2(1):173-181.

11. Abankwah V, Aidoo R, Osei RK. Socio-economic impact of government spraying programme on cocoa farmers in Ghana. J Sustain Dev Afr. 2010;12(4):116-26.

12. Acebo-Guerrero Y, Hernández-Rodríguez, A, Heydrich-Pérez, M, El Jaziri, M, Hernández-Lauzardo, A. N. Management of black pod rot in cacao (*Theobroma cacao* L.): A Review. Fruits. 2012;67(0):41-48.

13. World Bank. Supply Chain Risk Assessment: Cocoa in Ghana. Agricultural Risk Management Team of the Agricultural and Rural Development; 2008. Accessed 22 June 2013.
    Available:http://www.agriskmanagementforum.pdf

14. Baah F, Anchirinah V. A review of Cocoa Research Institute of Ghana extension activities and the management of cocoa

pests and diseases in Ghana. Am J Soc Mgmt Sci. 2011;2(1):196-201.

15. Ntiamoah A, Afrane G. Environmental impacts of cocoa production and processing in Ghana: Life cycle assessment approach. Journal of Cleaner Production. 2008;16(16):1735-40.

16. International Cocoa Organisation (ICCO). ICCO Quarterly Bulletin of Cocoa Statistics, Cocoa year. 2011/12. 2012;38(4). Accessed August 8 2013. Available:http://www.icco.org/ international-cocoa agreements/docdownload

17. Binam JN, Gockowski J, Nkamleu GB. Technical efficiency and productivity potential of cocoa farmers In West African Countries. The Developing Economies. 2008;46(3):242-63.

18. Ghana Statistical Service (GSS). Ghana living standards survey: Report of the Fifth round. Accra: Ghana; 2008. Accessed Aug 08 2013.
Available:http://www.statsghana.gov.gh/docfiles/glss5_report.pdf

19. Boon E, Ahenkan A, Domfeh KA. Human and environmental health linkages in Ghana: A case study of Bibiani-Bekwai and Sefwi Wiawso Districts. GRIN Verlag. Boon E, Ahenkan A, Domfeh KA. Human and environmental health linkages in Ghana: A case study of Bibiani-Bekwai and Sefwi Wiawso Districts. GRIN Verlag; 2008.

20. Bibiani-Anhwiaso-Bekwai District Assembly (BABDA). Medium Term Development Plan; 2006. Accessed June 12 2013.
Available:www.ghanadistricts.gov.gh/

21. Haidt J. The righteous mind: why good people are divided by politics and religion. London: Penguin; 2012.

22. Copestake J. Credible impact evaluation in complex contexts: Confirmatory and exploratory approaches. Evaluation. 2014;20(4):412-427.

23. Bosompem M, Mensah E. Occupational hazards among cocoa farmers in the Birim South District in the Eastern Region of Ghana. Journal of Agricultural and Biological Science. 2012;7(12):1055-1061.

24. Ministry of Food and Agriculture (MoFA). Districts-Western Region: Bibiani-Anhwiaso-Bekwai; 2013. Accessed Mar 07 2013. Available: http://mofa.gov.gh/site

25. Ghana Statistical Service (GSS). 2010 Population and Housing Census. Final Results.Ghana Statistical Service. Accra; 2012. Accessed 3 June 2013. Available:http://www.statsghana.gov.gh/docfi/2010_population_and_housing_census_final_results.pdf

26. Quarshie E, Nyarko BJB, Serfor-Armah Y. Studies of the levels of some toxic elements in soil and tailings from Bibiani Mining Area of Ghana. Research Journal of Environmental and Earth Sciences. 2011;3(3):1-11.

27. Anang BT, Adusei K, Mintah E. Farmers' assessment of benefits and constraints of ghana's cocoa sector reform. Current Research Journal of Social Science. 2011;3(6):465-470.

28. Robinson S, Mendelson AL. A qualitative experiment research on mediated meaning construction using a hybrid approach. Journal of Mixed Methods Research. 2012;6(4):332-347.

29. Ghana Statistical Service. Patterns and trends of poverty in Ghana: 1991–2006. Accra: Ghana Statistical Service; 2007.

30. Danso-Abbeam G, Aidoo R, Agyemang KO, Ohene-Yankyera K. Technical efficiency in Ghana's cocoa industry: Evidence from Bibiani-Anhwiaso-Bekwai District. Journal of Development and Agricultural Economics. 2012;4(10):287-94.

31. Aneani F, Anchirinah VM, Owusu-Ansah F, Asamoah M. Adoption of some cocoa production technologies by cocoa farmers in Ghana. Sustainable Agriculture Research. 2012;1(1):103-117.

32. Acheampong K, Addo, G, Baah, F, Bhar, R, Branch AC, Swaithes A. Hadley P, AJ. Daymond. Mapping cocoa productivity in Ghana. In: Proceedings of the 17th International Cocoa Research Conference. Yaoundé, Cameroon, October 2012. In press; 2014.

33. Asamoah M, Ansah FO, Anchirinah V, Aneani F, Agyapong D. Insight into the standard of living of Ghanaian Cocoa Farmers. Greener Journal of Agricultural Sciences. 2013;3(5);363-370.

34. Baffoe-Asare R, Danquah JA, Annor-Frempong F. Socioeconomic Factors Influencing Adoption of CODAPEC and Cocoa High-Tech 1 technologies among Smallholder Farmers in Central Region of Ghana 2. American Journal of Experimental Agriculture. 2013;3(2):277-92.

35. Adu-Ampomah Y, Asante EG, Opoku SY. Farmers' knowledge, attitudes, and

perceptions of innovation in cocoa production and implications for participatory improved germplasm development. In: International Workshop on Cocoa Breeding for Farmers' Needs. 2009. Ingenic and Catie, San José, Costa Rica; 2006.

36. Knudsen MH, Fold N. Land distribution and acquisition practices in Ghana's cocoa frontier: The impact of a state-regulated marketing system. Land Use Policy. 2011;28(1):378-387.

37. Helmsing AHJBHJ, Vellema S. Value Chains, Social Inclusion and Economic Development: Contrasting Theories and Realities. Taylor & Francis; 2012.

38. Teal F, Zeitlin A, Maamah H. Ghana Cocoa Farmers Survey. Report of the Ghana Cocoa Board. 2004. Centre for the Study of African Economies, Oxford University; 2006.

39. Opoku IY, Gyasi, E.K, Onyinah, G.K, Opoku, E, Fofie, T. The National Cocoa Diseases and Pests Control (CODAPEC) Programme: Achievements and Challenges. 2006. Proceedings of the 15th International Cocoa Research Conference. San Jose, Costa Rica. 2006;1007–1013.

40. Daniel R, Konam J, Saul-Maora J, Kamuso A, Namaliu Y, Vano JT, et al. Knowledge through participation: The triumphs and challenges of transferring Integrated Pest and Disease Management (IPDM) technology to cocoa farmers in Papua New Guinea. Food Sec. 2011;3(1):65-79.

41. Yaro JA. Is deagrarianisation real? A study of livelihood activities in rural northern Ghana. The Journal of Modern African Studies. 2006;44(01):125-56.

42. Owusu V, Abdulai A, Abdul-Rahman S. Non-farm work and food security among farm households in Northern Ghana. Food Policy. 2011;36(2):108-18.

43. Hadley C, Stevenson, EGJ, Tadesse Y, Belachew T. Rapidly rising food prices and the experience of food insecurity in urban Ethiopia: Impacts on health and well-being. Social Science and Medicine. 2012;75(1):2412-2419.

44. Headey B. Poverty Is Low Consumption and Low Wealth, Not Just Low Income. Soc Indic Res. 2008;89(1):23-39.

45. Coleman S. Where Does the Axe Fall? Inflation Dynamics and Poverty Rates: Regional and Sectoral Evidence for Ghana. World Development. 2012;40(12):2454-67.

46. Bogetic Y, Bussolo, M, Ye X, Medvedev D, Wodon Q, Boakye D. Ghana's growth story: how to accelerate growth and achieve MDGs? 2007. Background Paper for Ghana Country EM. April World Bank, Washington DC; 2007.

47. Breisinger C, Diao X. Economic transformation in theory and practice: what are the messages for Africa? IFPRI Discussion Paper 797. International Food Policy Research Institute, Washington D.C; 2008.

48. Breisinger C, Diao X, Thurlow J. Modelling growth options and structural change to reach middle income country status: The case of Ghana. Economic Modelling. 2009;26(2);514-525.

49. Coulombe H, McKay A. The estimation of components of household INCOMES and expenditures: A Methodological Guide Based on the Last Three Rounds of the Ghana Living Standards Survey, 1991/1992, 1998/1999 and 2005/2006.2008. Accessed July 07 2013. Available:http://personal.psc.isr.umich.edu/ ~davidl/GhanaCourse/GLSS/G5Aggregate %20Paper.pdf

50. Annim SK, Mariwah S, Sebu J. Spatial inequality and household poverty in Ghana. Economic Systems. 2012;36(1): 487-505.

51. Coulombe, H, Wodon, Q. 'Poverty Livelihoods and Access to Basic Services in Ghana', Ghana CEM: Meeting the Challenge of Accelerated and Shared Growth; 2007. Accessed 25 July 2013. Available:http://siteresources.worldbank.or g/ poverty.pdf

52. Alkire S, Foster JE. Counting and Multidimensional Poverty Measurement. Journal of Public Economics. 2011;95(1): 476-487.

53. Hargreaves JR, Morison LA, Gear JS, Makhubele MB, Porter JD, Busza J, Pronyk PM. Hearing the voices of the poor: Assigning poverty lines on the basis of local perceptions of poverty. A Quantitative Analysis of Qualitative Data from Participatory Wealth Ranking in Rural South Africa. World Development. 2007;35(2):212-229.

54. Waglé UR. Multidimensional poverty: An alternative measurement approach for the United States? Social Science Research. 2008;37(2):559-580.

55. Whelan CT, Layte R, Maître B. Understanding the mismatch between

income poverty and deprivation: A dynamic comparative analysis. European Sociological Review. 2004;20(4):287-302.

56. Mitra S, Jones K, Vick B, Brown D, McGinn E, Alexander M. Implementing a Multidimensional Poverty Measure Using Mixed Methods and a Participatory Framework. Soc Indic Res. 2013;110(3):1061-81.

57. Bhorat H, van der Westhuizen C. Non-monetary dimensions of well-being in South Africa, 1993–2004: A post-apartheid dividend? Development Southern Africa. 2013;30(3):295-314.

# 11

# Effect of Integrated Weed Management Practices on Weeds Infestation, Yield Components and Yield of Cowpea [*Vigna unguiculata* (L.) Walp.] in Eastern Wollo, Northern Ethiopia

Getachew Mekonnen[1,2*], J. J. Sharma[2], Lisanework Negatu[2] and Tamado Tana[2]

[1]*Department of Plant Sciences, College of Agriculture and Natural Resources, Mizan Tepi University, P.O.Box 260,Mizan Teferi, Ethiopia.*
[2]*School of Plant Sciences, College of Agriculture and Environmental Sciences, Haramaya University, P.O.Box 138, Dire Dawa, Ethiopia.*

***Authors' contributions***

*This work was carried out in collaboration between all authors. Author GM is PhD student designed the study, performed the statistical analysis, wrote the protocol and write first draft of the manuscript. Co-authors JJS, LN and TT are advisors and giving relentless guidance, valuable comments on the dissertation manuscript and encouragement during the course of research work and prepared the final manuscript for publication. All authors read and approved the final manuscript.*

*Editor(s):*
(1) Rusu Teodor, Department of Technical and Soil Sciences, University of Agricultural Sciences and Veterinary Medicine Cluj-Napoca, Romania.
(2) Mirza Hasanuzzaman, Department of Agronomy, Sher-e-Bangla Agricultural University, Bangladesh.
(3) Mintesinot Jiru, Department of Natural Sciences, Coppin State University, Baltimore, USA.
*Reviewers:*
(1) Anonymous, Brazil.
(2) Anonymous, Nigeria.
(3) Anonymous, Brazil.

## ABSTRACT

Cowpea [*Vigna unguiculata* (L.) Walp.] is usually infested and its yield is adversely affected by a number of weed species that compete with the crop from germination to harvest, affecting the crop yield adversely. Therefore, an experiment was conducted at Sirinka and Jari, northern Ethiopia during the 2013 main cropping season (July-October). The objectives were to assess the effect of pre-emergence s-metolachlor and pendimethalin on weeds, and growth, yield components and

---

*Corresponding author: E-mail: sibuhmekdes@gmail.com;*

yield of cowpea and to investigate the possibilities of supplementing low doses of herbicides with hand weeding for effective and cost effective weed management. There were 12 treatments comprising: s-metolachlor (1.0, 1.5 and 2.0 kg ha$^{-1}$); pendimethalin (1.0, 1.3 and 1.6 kg ha$^{-1}$), s-metolachlor at 1.0 kg ha$^{-1}$ + hand-weeding at 5 weeks after crop emergence (WAE), pendimethalin at 1.0 kg ha$^{-1}$ + handweeding at 5 WAE, one handweeding at 2 WAE, two handweeding at 2 and 5 WAE, weed free and weedy checks. The treatments were arranged in randomized complete block design with three replications. 78.6% of the weeds comprised in the experimental sites were the broadleaved. At 20 DAE, application of 2.0 kg ha$^{-1}$ s-metolachlor at both locations resulted in the lowest broadleaved weeds, sedge and total weed density. Pendimethalin failed to control *Commelina benghalensis* and *Xanthium strumarium*. At 55 DAE, low rate of s-metolachlor and pendimethalin when superimposed with one hand weeding were as effective as complete weed free treatment in reducing the broadleaved weeds and sedge density. The minimum weed dry weight was registered with the application 2.0 kg ha$^{-1}$ of s-metolachlor in both locations; however, at 55 days and harvest, weeds accumulated significantly lower dry weight due to1.0 kg ha$^{-1}$ s-metolachlor 1.0 kg ha$^{-1}$ pendimethalin superimposed with hand weeding at both locations. The interaction of location with weed management practices was significant on days to 50% flowering and physiological maturity of the crop, number of pods plant$^{-1}$, grain and aboveground dry biomass yield and yield loss. The maximum grain yield (4277 kg ha$^{-1}$) was obtained in complete weed free treatment at Sirinka which was statistically equivalent with complete weed free and two hand weeding treatments at Jari and Sirinka experimental sites respectively. Due to weed infestation throughout the crop growth, the highest yield loss (70.8%) was recorded at Jari while it was 47.5% at Sirinka. The highest gross benefit was obtained with the application of 1.0 kg ha$^{-1}$ of s-metolachlor superimposed with hand weeding followed by two hand-weeding at 2 and 5 WAE. Therefore, managing the weeds with the application of 1.0 kg ha$^{-1}$ of s- metolachlor + hand weeding and hoeing 35 DAE proved to be the most profitable practice. However, under the condition of labour constraint and timely availability of the herbicide, pre emergence application of 2.0 kg ha$^{-1}$ of s-metolachlor should be used to preclude the yield loss and to ensure maximum benefits.

Keywords: *Broadleaved and grass weeds; economic analyses; herbicides; yield loss.*

## 1. INTRODUCTION

Cowpea is of the most important crop to the livelihoods of millions of relatively poor people in less developed countries of the tropics [1]. It is extensively grown in the lowlands and mid-altitude regions of Africa, sometimes as sole crop but more often intercropped with cereals such as sorghum or millet [2]. It is a good food security crop as it mixes well with other recipe [3]. Cowpea fixes atmospheric nitrogen through symbiosis with nodule bacteria [4]. It does well and is most popular in the semi-arid of the tropics where other food legumes do not perform well [5].

A number of weed species are affecting the yield by competing with the crop from germination to harvest [6], and this yield loss in cowpea which ranged from 12.7% - 60.0% is due to weeds [7]. According to [8], the presence of weeds in cowpea reduced yield by 82% and a significant increase in yield of pods was noted by controlling weeds up to 45 days of sowing. Therefore, in order to enhance crop yield, weed control during this period is very important. The physical and mechanical approaches of weed control are very expensive as labour is usually unavailable during the peak periods of weed removal from the field [9]. Hand weeding required over 50% of the farmers' time leaving them with little or no time for other activities [10]. In this regard, the use of herbicides to control weeds in cowpea fields appears to be the other option [11]. Herbicide use would improve the lives of farmers by eliminating the need for back-breaking labour.

Significantly higher grain yield and net return of cowpea were obtained with pendimethalin applied pre-emergence at 0.75 kg ha$^{-1}$+ hand-weeding at 5 weeks after planting (WAP) compared to other treatments [6,12] reported that pendimethalin at 1.0 kg ha$^{-1}$+ hand weeding at 30 days after planting significantly gave a higher cowpea grain yield, weed density and biomass were the lowest in this treatment. Metolachlor has an excellent action against annual grasses and *Cyperus* species. Research with metolachlor in cowpeas resulted in yields comparable to those receiving the recommended two weeding [13]. However, the rate of s-metolachlor may depend upon soil types, rainfall and irrigation patterns, temperature, crops and weeds; nevertheless, 1.5 kg ha⁻¹of s-metolachlor has been used in pulse crops in Ethiopia [14]. Use of herbicides may therefore provide a timely and adequate alternative to hand weeding as this not only removes the drudgery associated with it but also lowers the cost of weeding and provides protection for crop against early weed competition when pre-emergence herbicides are used [10].

Integrating herbicides with cultural methods is an option for better weed control. Integrated weed management (IWM) does not preclude herbicide use, it includes their judicious use along with other agronomic methods that help crops compete with weeds and reduce weed seed production. IWM involves using an agronomical approach to minimize the overall impact of weeds and, indeed, maximize the benefits.

S-metolachlor and pendimethalin which are among the recently introduced herbicides in Ethiopia have not been widely evaluated in cowpea specifically in the study area. Hence, the objectives of this study were to assess the effect of s-metolachlor and pendimethalin on weeds, growth, yield components and yield of cowpea. It

was also meant to investigate the possibilities of supplementing low doses of herbicides with hand weeding for effective weed control and their economic returns in cowpea.

## 2. MATERIALS AND METHODS

### 2.1 Description of the Study Area

The experiment was conducted at Sirinka Agricultural Research Center experimental sites at Jari (11°21'N latitude and 39°38'E longitude; 1680 m. a. s. l. altitude) and Sirinka (11°45'00" N latitude; 39°36'36"E longitude; 1850 m. a .s .l. altitude) in northern Ethiopia during the 2013 main cropping season (July to October ).The soil of the experimental fields was clay loam and clay with the pH of 6.95 and 6.91 at Sirinka and Jari, respectively. At Sirinka the organic carbon was 1.37%, total N was 0.09%, available P 12.17 mg kg⁻¹ soil and CEC 53.44 cmol$_c$ kg⁻¹ while respective values at Jari were 1.33%, 0.07%, 9.17 mg kg⁻¹ and 33.44 cmol$_c$ kg⁻¹. The total seasonal rainfall received during the crop season was 750.4 mm and 589.1 mm at Sirinka and Jari with mean maximum and minimum temperatures of 28.6 and 14.7°C, and 29.6 and15.8°C, respectively (Fig. 1).

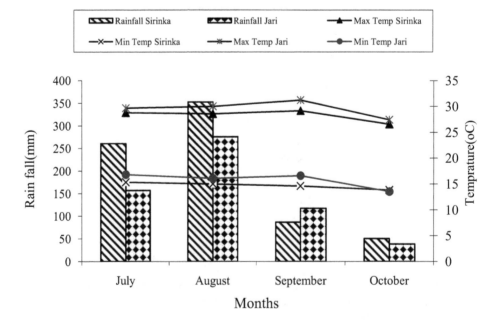

**Fig. 1. Monthly mean maximum and minimum temperatures (°C) and total rainfall (mm) at Jari and Sirinka in 2013 main cropping season**
*Source: Sirinka Agricultural Research Center*

## 2.2 Treatments and Experimental Design

The experiment of this study comprises of 12 treatments: s-metolachlor at (1.0, 1.5 and 2.0 kg ha$^{-1}$), pendimethalin (1.0, 1.3 and, 1.6 kg ha$^{-1}$), s-metolachlor at 1.0 kg ha$^{-1}$ + hand weeding at 5 weeks after crop emergence (WAE), pendimethalin at 1.0 kg ha$^{-1}$ + hand weeding at 5 WAE, one hand weeding at 2 WAE, two hand weeding at 2 and 5 WAE, weed free check and weedy check. The design of the experiment was randomized complete block design with three replications.

The plot size was 3.6 m x 2.4 m. The cowpea variety Asrat (IT 92KD-279-3) which is bush and trailing type I was planted at inter- and intra- row spacing of 60 cm and 10 cm, onthe 8$^{th}$July and 13$^{th}$ of July, 2013 at Jari and Sirinka, respectively. Fertilizer (100 kg DAP; 18 kg N+46 kg P$_2$O$_5$ ha$^{-1}$) was applied uniformly to each plot at the time of sowing. The pre-emergence herbicides were applied at the specific rates using Knapsack sprayer one day after planting using flat-fan nozzle. The spray volume was 450 l ha$^{-1}$. The outermost one row from each side and 3 plants on each end of rows were excluded to remove border effect. Thus, the net plot area was 2.4 m x 1.8 m. The crop was harvested on October 15 and 25, 2013 at Jari and Sirinka, respectively.

## 2.3 Data Collection and Analysis

Data on weed flora present in the experimental fields were recorded during the experimental period. The weed density was recorded by throwing a quadrate (0.25 m×0.25 m) randomly at two places in each plot at about 15 days before the expected harvest time. The weed species found within the sample quadrat were identified, counted and expressed in m$^{-2}$. For the aboveground weed dry weight/ biomass, the weeds falling within the quadrate were cut near the soil surface immediately after taking observation on weed count and placed into paper bags separately treatment wise. The samples were sun dried for 3-4 days and thereafter were placed in to an oven at 65°C temperature till their constant weight and subsequent dry weight was measured. The dry weight was expressed in g m$^{-2}$.

Weed Index: It was measured from a particular treatment when compared with a weed free treatment and expressed as percentage of yield potential under weed free.

Weed Index= $\frac{x-y}{x}X100$

Where

x= Yield from weed free check; y= Yield from a particular treatment

Weed Control Efficiency (WCE): was calculated using the following formula

$$\text{WCE} = \frac{WDC - WDT}{WDC} \times 100$$

Where WCE= Weed Control Efficiency, WDC=Weed dry weight in weedy check, and WDT= Weed dry weight in a particular treatment

Number of days to 50% flowering was recorded as number of days from emergence of cowpea to the date when first flower appeared on 50% of the plants in each plot, whereas days to maturity was recorded as the number of days from planting to the day when 90% of the plants reached physiological maturity, *i.e.* both pods and plants turned yellow (senescing) based on visual observation. Plant height (cm) was taken with a measuring tape from 10 randomly selected and pre tagged plants in each net plot area from the base to the apex of the main stem at physiological maturity. The number of pods plant$^{-1}$ was taken from the total pods of the above tagged plants at harvest. The total number of seeds from the above pods was taken and counted to average the number of seeds pod$^{-1}$. Out of seeds from the above, 100 seeds were counted and their weight was recorded at 10.5% moisture content for hundred seed weight (g). Harvest index (%) was determined by harvesting ten plants in each plot at physiological maturity and their dried aboveground biomass was recorded and then as grain yield divided by the aboveground dry biomass. Treatment per plant dry weight of straw was multiplied by the number of plants in respective treatments. This was considered as the aboveground dry biomass weight. The grain weight obtained in ten plants was added to the final yield. The grain yield (kg ha$^{-1}$) was measured after threshing the sun dried plants harvested from each net plot and the yield was adjusted at 10.5% seed moisture content.

Data on weed community, weed density, weed dry biomass; crop phenology, growth, yield attributes and yield were subjected to analysis of variance using GenStat 15.0 computer software [15]. Fisher's protected Least Significant Difference (LSD) test at 5% level of significance was used to separate the differences among treatment means (P < 0.05) [16]. As the F-test of the error variances for the parameters of the two sites was homogeneous, combined analysis of data was used.

## 2.4 Partial Budget Analysis

The concepts used in the partial budget analysis were the mean grain yield under each treatment for both locations, the field price of the crop (sale price minus the costs of harvesting, threshing, winnowing, bagging and transportation), the varied total costs including the sum of field cost of herbicide and its application. Actual yield was adjusted downward by 10% to represent the difference between the experimental yield and the yield farmers could get from the same treatment [17].

## 3. RESULTS AND DISCUSSION

### 3.1 Effect of Weed Management Practices on Weeds

#### 3.1.1 Weed community

The major weeds in the experimental fields were broadleaved, while sedges were found at lesser extent. There was only one grass species present at Sirinka to a very limited extent. Hence this weed was merged with sedges for the purpose of describing the results. The parasitic weed broomrape (*O. cerenata*) was found at Jari in plots infested with *X. strumarium* only. The remaining weeds were found at both locations. The weed flora present in the experimental fields is presented in Table 2.

**Table 1. Description of herbicides used in the experiment**

| Common name | Trade name | Chemical name |
|---|---|---|
| S-metolachlor | Dual Gold 960EC | [2-chloro-6`-ethyl-N-(2-methoxy-1-methylethyl)acet-o-toluidide] |
| Pendimethalin | Stomp Extra 38.7% CS | [N-(1-ethylpropyl)-2, 6-dinitro-3, 4-xylidine] |

#### 3.1.2 Weed density

*3.1.2.1 Weed density at 20 days after crop emergence*

Weed density showed a significant difference (P <0.01) due to various weed management practices. At 20 DAE, the application 2.0 kg ha$^{-1}$ of s-metolachlor at Jari and Sirinka resulted in the existence of lowest broadleaved weed density. Furthermore, at Sirinka, no significant differences existed between s-metolachlor at 1.5 kg ha$^{-1}$, pendimethalin 1.0 kg ha$^{-1}$ + handweeding at 35DAE, and pendimethalin at 1.0 kg ha$^{-1}$. The density of broadleaved weeds decreased with the increase in s- metolachlor application rates but significant difference was notobserved between 1.0 and 1.5 kg ha$^{-1}$ rates (Table 2). This trend was not found for pendimethalin wherein pendimethalin at 1.0 kg ha$^{-1}$ recorded the minimum weed density which was significantly lower than pendimethalin at 1.3 kg ha$^{-1}$ but in parity with pendimethalin at 1.6 kg ha$^{-1}$. One hand-weeding at 2 WAE (or 14 DAE), two hand-weeding at 2 and 5 WAE ( or 14 and 35 DAE) and pendimethalin at 1.3 kg ha$^{-1}$ had higher weed density, but the weedy check plots showed appreciably highest broadleaved weeds density than the other weed management practices at both the locations.

At 20 DAE, the sedge density at Jari was minimum for s-metolachlor at 2.0 kg ha$^{-1}$ treated plots, which had no significant difference with its lower rates, and s-metolachlor at 1.0 kg ha$^{-1}$ + handweeding at 5 WAE (35 DAE).The application of s–metolachlor proved superior to pendimethalin in controlling the sedges; nevertheless, the performance of pendimethalin was significantly better than weedy check. At Sirinka the application of 2.0 kg ha$^{-1}$ s-metolachlor also resulted in the lowest sedge population and was statistically at par with s-metolachlor at 1.5 kg ha$^{-1}$, pendimethalin at 1.0 kg ha$^{-1}$, and low dose of both the herbicides superimposed with one hand-weeding at 5 WAE (35 DAE).

The total weed density was lowest with the application of 2.0 kg ha$^{-1}$ of s- metolachlor at Jari while at Sirinka, the lowest total weed density was obtained with s-metolachlor at 2.0 kg ha$^{-1}$, but it had no significant difference with s-metolachlor at 1.5 kg ha$^{-1}$, pendimethalin at 1.0 kg ha$^{-1}$ + handweeding at 5WAE (35 DAE)and pendimethalin at 1.0 kg ha$^{-1}$.The increasing s-metolachlor application rates decreased the total weed density but there was no significant difference observed between 1.0 and 1.5 kg ha$^{-1}$ treatments of s-metolachlor at Jari. In contrast, pendimethalin at 1.0 kg ha$^{-1}$ resulted insignificant decrease in total weed density over its higher rates. In the experimental field, it was observed that the application of pendimethalin failed to control *C. benghalensis* and at lower rate of its application, there might be more inter specific competition between the weeds thereby resulting in reduced overall weed density while under higher rates, the weeds other than C. benghalensis were also controlled which reduced inter specific competition. Further, this in turn might have provided greater opportunity to C. benghalensis to germinate in larger amount. At Sirinka the total weed density also decreased with the increase in s-metolachlor application rate, but at jari, there was no significant different observed between 1.5 and 2.0 kg ha$^{-1}$rates. The effect of pendimethalin on total weed density at Sirinka was similar to that of Jari. The significant reduction in weed density with lowest pendimethalin application rate at both locations was in contrast to the findings of [17] who stated that reduced rates of herbicide are not advisable under heavy weed pressure. However, it seemed that the weed species and their composition also determined the effectiveness of the herbicide. At both locations, the total weed density was significantly higher in weedy check than the other weed management practices; however, the results depicted higher weed pressure at Jari than at Sirinka.

### 3.1.2.2 Weed density at 55 days after crop emergence

At 55 DAE, the density of broadleaved weeds was lowest due to the application of 1.0 kg ha$^{-1}$of pendimethalin superimposed with one hand-weeding at 5 WAE at Jari. However, it had no significant difference with s-metolachlor at 1.0 kg ha$^{-1}$ treatment combined with one hand-weeding at 5 WAE and two hand-weeding at 2 and 5 WAE. The results also depicted that low rate of s-metolachlor and pendimethalin when combined with one hand-weeding were as effective as complete weed free treatment to reduce the broadleaved weed density. Alike what was at 20 DAE, the broadleaved weed density also decreased with the increase in s- metolachlor application rates whereby 1.5 and 2.0 kg ha$^{-1}$rates did not significantly differ weed density, but there observed a significant reduction in density over 1.0 kg ha$^{-1}$rate. In case of pendimethalin, the results remained inconsistent. Two hand-weeding proved significantly better than one hand-weeding in reducing the broadleaved weed density. This might partially be due to the late emerging X. strumarium which infested the plots after one hand-weeding was resorted. On the other hand, one hand-weeding was found to bring significant reduction in broadleaved weed density at all rates of pendimethalin treatments.

At Sirinka, the application of 1.0 kg ha$^{-1}$of s-metolachlor + one hand-weeding at 5 WAE gave the lowest weed density which was not significantly different with two hand-weeding. Both of these practices were found to be significantly better than other weed management practices. The application of higher rates (1.5 and 2.0 kgha$^{-1}$) of s-metolachlor gave significant reduction in density over its lower application rates while at higher rates (1.3 and 1.6 kgha$^{-1}$) of pendimethalin it significantly increased over the lower rates. Furthermore, the poor control of these weeds with herbicide alone might be due their bigger seed size which enabled them to emerge from deeper soil layer as the weeds emerging from deeper layers (herbicide free zone) are selective to be applied herbicides due to positional selectivity.

Late emerging weed X. strumarium was not controlled by both of these herbicides which may be due to bigger seed size that enabled it to germinate from deeper soil depth in herbicide free zone thus escaping the herbicide interference in the germination process. On the other hand, pendimethalin treatment failed to control C. benghalensis. However, when the low dose of herbicides was superimposed with hand-weeding at 5 WAE, it might have contained the infestation by these weeds resulting in significantly lower broadleaved weed density than the herbicides alone. Moreover, the infestation of X. strumarium which might have contributed to higher weed density was more persistent at Jari than at Sirinka.

The sedges population was lowest in pendimethalin at 1.0 kg ha$^{-1}$+ one hand-weeding at 5 WAE at Jari which did not significantly differ with s-metolachlor at 1.0 kg ha$^{-1}$ combined with one hand-weeding at 5 WAE, s- metolachlor at 2.0 kg ha$^{-1}$and two hand-weeding at 2 and 5 WAE. All these practices were statistically at par with complete weed free. At Sirinka, low dose of both s-metolachlor and pendimethalin each superimposed with one handweeding resulted in the lowest sedges population which had no significant difference with s-metolachlor at 2.0 kg ha$^{-1}$, one and two hand-weeding. At Jari and Sirinka, weedy check had significantly higher density than the other weed management practices.

At 55 DAE, the total weed density was lowest with the application of pendimethalin at 1.0 kg ha$^{-1}$+ one hand-weeding, however, it had no significant difference with s-metolachlor at 1.0 kg ha[-1]+ one hand-weeding and two hand-weeding at 2 and 5 WAE at Jari. At Sirinka, the application of 1.0 kg ha[-1]of s-metolachlor at + one hand weeding registered the minimum total weed density which was significantly lower than other weed management practices. Though, at Jari there was no significant difference in total weed density found between pendimethalin at 1.0 kg ha[-1]+ one hand weeding , s-metolachlor at 1.0 kg ha[-1]+ one hand weeding and complete weed free, but at Sirinka, significant a variation was observed between these treatments (Table 2). In total weed density, the trend due to s-metolachlor and pendimethalin application was similar to that of broadleaved weeds and sedges at both sites.

### 3.1.2.3 Weed density at harvest

The broadleaved weed density at crop harvest obtained due to application of 1.0 kg ha[-1]of s-metolachlor + hand weeding at 5 WAE, 1.0 kg ha[-1] of pendimethalin + hand weeding at 5 WAE, 2.0 kg ha[-1]of s-metolachlor and two hand weeding at 2 and 5 WAE was statistically in parity with complete weed free and resulted in significantly lower density than the other weed management practices at Jari (Table 3). Also, at Sirinka a similar trend was observed but no significant difference was found with one hand weeding at 2 WAE. The effect of weed management practices on sedges density was similar to broadleaved weeds at Jari; however, the application of 1.5 kg ha[-1]s- metolachlor at was also in statistical parity.

The application of s–metolachlor alone, 1.0 kg ha[-1]of pendimethalin, low dose of these herbicides combined with hand weeding and hand weeding treatments were statistically at par with each other and complete weed free except s-metolachlor and pendimethalin each at 1.0 kg ha[-1]at Sirinka. All these weed management practices significantly reduced sedge density over other treatments. The density of both broadleaved weeds and sedges were lower at crop harvest than at 55 DAE. This might be due to the competitive effect of the crop especially for the solar radiation. In line with this result, [18] described that plants with large leaf area indices have a competitive advantage and normally out-compete plants with smaller leaf areas.

The total weed density was significantly reduced with the application of s-metolachlor at 1.0 kg ha[-1]+ hand weeding at 5 WAE, pendimethalin at 1.0 kg ha[-1] + hand weeding at 5 WAE, s-metolachlor at 2.0 kg ha[-1] and two hand weeding at 2 and 5 WAE over other weed management practices at Jari whereas, one hand weeding had also similar effect at Sirinka experimental site. All these practices were statistically equivalent with complete weed free in the practice of reducing total weed density (Table 3). Hand weeding uprooted the emerged weeds which were in turn suppressed by the crop canopy that brings about decreased weed density at crop harvest. The weedy check plots resulted in significantly more total weed density than all other weed management practices that could be attributed to unchecked growth of early and late emerging weeds. The application of herbicide or hand weeding, however, caused mortality of weeds causing lower weed density at harvest. [19,20,9] have also reported a maximum weed density in weedy check and weed control methods like application of herbicides and hand weeding and hoeing significantly deceased weed density over weedy check.

**Table 2. Weed community recorded in cowpea, at the experimental sites of Jari and Sirinka in main cropping season of 2013**

| Weed species | Family | Life form (category) |
| --- | --- | --- |
| Amaranthus spinosus L. | Amaranthaceae | Annual (broadleaved) |
| Amaranthus hybridus L. | Amaranthaceae | Annual (broadleaved) |
| Bidens pilosa L. | Asteraceae | Annual (broadleaved) |
| Commelina benghalensis L. | Commelinaceae | Annual (broadleaved) |
| Cyperus esculentus L. | Cyperaceae | Annual (sedge) |
| C. rotundus L. | Cyperaceae | Perennial (sedge) |
| Datura stramonium L. | Solanaceae | Annual (broadleaved) |
| Galinsoga parviflora Cav. | Asteraceae | Annual (broadleaved) |
| Orobanche crenata Forsk. | Orobanchaceae | Annual (broadleaved) |
| Oxalis latifolia Kunth. | Oxalidaceae | Annual (broadleaved) |
| Seteria verticillata (L.) Beauv. | Poaceae | Annual (grass) |
| Solanum nigrum L. | Solanaceae | Annual (broadleaved) |
| Tagetes minuta L. | Asteraceae | Annual (broadleaved) |
| Xanthium strumarium L. | Asteraceae | Annual (broadleaved) |

Table 3. Effect of weed management practices in cowpea on weed density (m⁻²) at 20 and 55 days after crop emergence (DAE) at Jari and Sirinka in 2013 main cropping season

| Weed management practices | Weed density ($M^{-2}$) | | | | | | | | | | | |
| --- | --- | --- | --- | --- | --- | --- | --- | --- | --- | --- | --- | --- |
| | 20 DAE | | | | | | 55 DAE | | | | | |
| | Broadleaved | | Sedges | | Total | | Broadleaved | | Sedges | | Total | |
| | Jari | Sirinka | Jari | Sirinka | Jari | Sirinka | Jari | Sirinka | Jari | Sirinka | Jari | Sirinka |
| S-metolachlor at 1.0 kg ha⁻¹ | 112.0[d] | 50.67[cd] | 13.33[cde] | 10.67[de] | 125.3[d] | 61.33[d] | 90.7[d] | 58.67[d] | 26.67[e] | 21.33[cd] | 117.3[e] | 80.00[d] |
| S-metolachlor at1.5 kg ha⁻¹ | 82.7[d] | 32.00[e] | 8.00[def] | 8.00[ef] | 90.7[d] | 40.00[e] | 50.7[e] | 34.67[ef] | 10.67[fg] | 16.00[de] | 61.3[f] | 50.67[e] |
| S-metolachlor at 2.0 kg ha⁻¹ | 40.0[e] | 26.67[e] | 2.67[ef] | 5.33[ef] | 42.7[e] | 32.00[e] | 34.7[ef] | 40.00[e] | 5.33[fg] | 13.33[ef] | 40.0[fg] | 53.33[e] |
| Pendimethalin at1.0 kg ha⁻¹ | 152.0[c] | 34.67[e] | 21.33[c] | 8.00[de] | 173.3[c] | 42.67[e] | 192.0[c] | 85.33[c] | 58.67[c] | 26.67[c] | 250.7[c] | 112.00[c] |
| Pendimethalin at1.3 kg ha⁻¹ | 192.0[ab] | 61.33[abc] | 66.67[a] | 13.33[cd] | 258.7[ab] | 74.67[c] | 224.0[b] | 106.67[b] | 72.00[b] | 37.33[b] | 296.0[b] | 144.00[b] |
| Pendimethalin at 1.6 kg ha⁻¹ | 176.0[bc] | 66.67[ab] | 53.33[b] | 24.00[ab] | 229.3[b] | 90.67[ab] | 181.3[c] | 101.33[b] | 53.33[c] | 37.33[b] | 234.7[c] | 138.67[b] |
| S-metolachlor at 1.0 kg ha⁻¹ + hand weeding at 5 WAE | 109.3[d] | 48.00[d] | 8.00[def] | 8.00[de] | 117.3[d] | 56.00[d] | 10.7[gh] | 18.67[g] | 2.67[g] | 8.00[f] | 13.3[hi] | 26.67[g] |
| Pendimethalin at 1.0 kg ha⁻¹ + hand weeding at 5 WAE | 149.3[c] | 32.00[e] | 18.67[cd] | 8.00[de] | 168.0[c] | 40.00[e] | 8.0[gh] | 29.33[f] | 0.00[g] | 8.00[f] | 8.0[hi] | 37.33[f] |
| One hand weeding at 2 WAE | 210.7[a] | 58.67[bcd] | 64.00[ab] | 18.67[bc] | 274.7[a] | 77.33[c] | 109.3[d] | 42.67[e] | 37.33[d] | 10.67[ef] | 146.7[d] | 53.33[e] |
| Hand weeding at 2 and 5WAE | 208.0[ab] | 61.33[abc] | 61.33[ab] | 21.33[ab] | 269.3[a] | 82.67[bc] | 24.0[fg] | 26.67[fg] | 5.33[fg] | 13.33[ef] | 29.3[gh] | 40.00[f] |
| Weed free check | 0.0[f] | 0.0[f] | 0.00[f] | 0.00[f] | 0.0[f] | 0.00[f] | 0.0[h] | 0.00[h] | 0.00[g] | 0.00[g] | 0.0[i] | 0.00[h] |
| Weedy check | 213.3[a] | 72.00[a] | 72.00[a] | 26.67[a] | 285.3[a] | 98.67[a] | 330.7[a] | 152.00[a] | 98.67[a] | 45.33[a] | 429.3[a] | 197.33[a] |
| LSD (5%) | 32.65 | 11.73 | 11.53 | 5.65 | 34.81 | 12.75 | 22.00 | 9.13 | 7.96 | 5.97 | 24.95 | 9.84 |
| CV (%) | 14.1 | 15.3 | 21.0 | 26.4 | 12.1 | 13.0 | 12.4 | 9.3 | 15.2 | 17.8 | 10.9 | 7.5 |

CV= coefficient of variation; LSD= least significant difference; WAE =weeks after emergence, Means in coloumns of same parameter followed by the same letter(s) are not significantly different at 5% level of significance

### 3.1.3 Weed dry matter weight

The influence of weed management practices at all the growth stages, in both the locations, on the weed dry matter weight was highly significant.

At 20 DAE, minimum weed dry weight was registered with the application of 2.0 kg ha$^{-1}$ of s-metolachlor at both locations (Table 3).Herbicide application at both places resulted in significant reduction in weed dry weight over weedy check. With the increase in S-metolachlor application rates, the weed dry weight significantly decreased, while the results were inconsistent with the application of pendimethalin at 20 and 55 DAEat both locations. At 55 DAE, weeds in plots treated with 1.0 kg ha$^{-1}$of s-metolachlor + hand weeding at 5 WAE and 1 kg ha$^{-1}$ of pendimethalin + hand weeding at 5 WAE at both locations accumulated significantly the lowest dry weight which might be due to the cumulative effect of herbicide and hand weeding (Table 3). [21] also obtained lower dry weight of weeds with 1.0 kg ha$^{-1}$ of butachlor at in combination with cultural practices, which was at par with weed free check. The results also revealed significant reduction in weed dry weight with two hand weeding as compared to one hand weeding and herbicides applied alone at this stage. The advantage of twice hand weeding over one hand weeding might be due to reduced soil seed bank as well as the weeds that emerged after second hand weeding were shorter in growth than the weeds that emerged after first hand weeding . Hand weeding controlled the emerged weeds and those that emerged later on might have failed to accumulate sufficient dry matter owing to the competition offered by the crop plants 55 DAE. Moreover, the weed seeds under depleted soil seed bank that might have been brought to the upper soil layer by hand weeding, though germinated and emerged later, but were in their initial growth stage thus accumulated less dry weight.

There was great difference in weed dry matter between the locations under respective weed management practices which might be the result of difference in weed density and, the environment. [22] also reported that herbicide application decreased the dry biomass of weeds; however, this decrement depends on several factors ,for example, duration of the crop, type of weed species, herbicides, fertilizer, etc. The rate of metolachlor application may depend upon soil types, rainfall and temperature. Similarly, [23] found that 1.5 kg ha$^{-1}$ of this herbicide is to be effective for the control of weeds in common bean. At crop harvest also similar effect of weed management practices was observed (Table 3). [9,24] also concluded that dry weight of weeds was significantly reduced in herbicide treated plots.

At 55 DAE weeds in plots treated with 1.0 kg ha$^{-1}$ of s-metolachlor + hand weeding at 5 WAE and 1.0 kg ha$^{-1}$ of pendimethalin + hand weeding at 5 WAE at both locations accumulated significantly the lowest dry weight due to the cumulative effect of herbicide and hand weeding (Table 3). At the time of crop harvest similar effect of weed management practices was observed (Table 3). [21] also obtained lower dry weight of weeds with 1.0 kg ha$^{-1}$of butachlor in combination with cultural practices, which was equivalent with weed free check. [9] and [24] also concluded that dry weight of weeds was significantly reduced in herbicide treated plots. The result of this study was in agreement with earlier works by [25,26] who observed reduced weed dry weight when herbicide application in common bean was combined with one hand weeding. Better control of weeds at the early stages by applying 1.0 kg ha$^{-1}$ of fluchloralin and subsequent removal of weeds by hand weeding at 40 DAE resulted in lesser weed count and weed dry weight [27].

Hand weeding controlled the emerged weeds and those weeds which would emerged later that might have failed to accumulate sufficient dry matter due to the competition offered by well grown crop plants. Further, the weed seeds under depleted soil seed bank that might have been brought to the upper soil layer by hand weeding, though germinated and emerged later, but were in their initial growth stage thus accumulated less dry weight.

At the time harvest, the weed dry weight accumulation with the increasing rates of s-metolachlor application significantly decreased at Jari, but no significant difference was obtained between the treatment of 1.5 and 2.0 kg ha$^{-1}$ of s-metolachlor at Sirinka. In contrast, the application of 1.3 and 1.6 kg ha$^{-1}$ pendimethalin had significantly higher weed dry weight than its lower rate at Jari, whilethe difference existed between these rates at Sirinkawas not significant. Moreover, the treatment with 1.0 kg ha$^{-1}$pendimethalin had weed dry weight statistically in parity with s-metolachlor rates at Sirinka (Table 3). The occurrence of significantly higher weed dry weight with increased pendimethalin application at Jari might be owing to the presence of the severe infestation with *X. strumarium.* On the other hand, at Sirinka, this weed was not found and the difference observed in weed dry weight was not significant. The data also depicted that the weedy check registered the highest weed dry weight which was significantly higher than other weed management practices.

Two hand weeding proved significantly better than one hand weeding in reducing weed dry weight at Jari whereas there was no significant difference observed at Sirinka. The contrasting results might be due to the extent to which the weed species and or the density differed at both locations. The results depicted that the application of s-metolachlor and pendimethalin at their lowest rates combined with one hand weeding provided prolonged weed control, and significant reduction in weed dry weight at harvest was observed like what was registered at earlier growth stages. Moreover, at Sirinka, s-metolachlor combined with hand weeding was as effective as complete weed free treatment in reducing the weed dry weight at the time of harvest. The effectiveness of both herbicides applied alone decreased with the increase in crop growth stage and this was more pronounced in case of pendimethalin. This might be due to late emerging weeds in herbicide treated plots that may be the consequence of loss of activeness of a herbicide (Table 3). At Jari experimental site, *X. strumarium* grew faster in the absence of inter-specific competition with other weed species especially in pendimethalin treated plots.

At both locations, weeds accumulated higher dry weight in weedy check plots and it was significantly higher than other weed management practices (Table 3). The higher weed dry weight in weedy check might be due to higher weed density that provided an opportunity to the weeds to compete vigorously for nutrients, space, light, water and carbon dioxide resulting in higher biomass production. Application of herbicides not only reduced the density of weeds but also suppressed the weed growth bringing about lower dry weight. These results are in agreement with the findings of [19,28] who reported maximum weed dry weight in weedy check. [29] reported that the weeds that germinated earlier or at the same time as the crop emergence, offered a serious competition as they got an opportunity to establish and accumulate dry matter weight faster than the crop.

## 3.2 Effect of Weed Management Practices on Crop Phenology and Growth Yield Attributes and Yield

### 3.2.1. Crop phenology and growth

*3.2.1.1 Days to 50% flowering*

Days to 50% flowering was significantly influenced by location, weed management practices and their interaction. It was observed that at Sirinka the application of s-metolachlor at 1.0 kg ha$^{-1}$+ hand weeding at 5 WAE resulted in significantly earlier flowering than Jari.

This was followed by the combination of pendimethalin at 1.0 kg ha$^{-1}$+ one hand weeding at 5 WAE at the same location which was statistically at par with the application of s-metolachlor at 1 kg ha$^{-1}$+ hand weeding at five WAE at Jari. The 50% flowering was delayed when weeds grown uninterrupted at both locations. However, this delay was significant at Sirinka. In conformity with this result, [30] also identified that the plants in unweeded plots took the highest time to reach 50% flowering. In general, application of either 1.0 kg ha$^{-1}$of s-metolachlor or 1.0 kg ha$^{-1}$ of pendimethalin each combined with one hand weeding at 5WAEleads to enhanced 50% flowering at both locations. This is consistent with the finding of [31], who stated that treating plots with chemical and supplementing with hand weeding at intervals helped to reduce the number of days to flowering and maturity.

*3.2.1.2 Days to 90% physiological maturity*

The effect of location, treatments and their interaction had a significant effect on 90% physiological maturity of the crop. The result within location treatments did not reveal significant difference in days to physiological maturity, however, at Sirinka; it was significantly delayed under all the treatments compared to what was found at Jari. The delayed maturity by about 10 days at Sirinka (Table 4) could be due the differences in amount and distribution of rain fall, temperature and elevation. The result was in contrast to the findings of [31] who stated that treating plots with chemical and supplementing with hand weeding at intervals helped to reduce number of days to maturity.

*3.2.1.3. Plant height*

The data (Table 5) showed that the plant height was significantly affected due to location while the weed management practices and its interaction with location had no significant effect.

The plants at Jari experimental site had significantly higher height by 12.3% than that of Sirinka. The higher temperature at Jari might have triggered growth resulting in increased plant height. More sunlight penetration to the crop plants also made photosynthates available, however, no significant difference in plant height was found between the weed management practices despite a great variation in weed density and dry weight (Table 2; Table 3). In contrast, [32] found differences in plant height due to various intensities of weed competition with crop plants.

### 3.2.3 Effect of weed management practices on yield attributes and yield

#### 3.2.3.1. Stand count at harvest

The stand count at harvest was significantly influenced by location and weed management practices, but their interaction had no significant effect (Table 6). The final crop stand was significantly higher by 8.1% at Sirinka than at Jari. The weed density as well as weed dry weight was higher at Jari than at Sirinka (Table 2; Table 3) might have contributed for the lower survival of crop plants. The highest stand count was recorded from the treatment of s-metolachlor at 1.0 kg ha$^{-1}$+ hand weeding at 5 WAE (120756 plants ha$^{-1}$) which was statistically in parity with weed free check, pendimethalin 1.0 kg ha$^{-1}$+ hand weeding at 5 WAE and two hand weeding at 2 and 5 WAE.

Comparatively higher survival of the plants observed under these weed management practices could be due to better weed control. The significantly lower plant stand under weedy check might be due to severe competition for growth resources particularly for space and light that suppressed crop plants the extent that the crop plants could not survive.

#### 3.2.3.2. Number of pods per plant

The location, weed management practices and their interaction had significant effect on number of pods plant $^{-1}$. The interaction effect revealed highest number of pods plant$^{-1}$ obtained with the application of s-metolachlor at 1.0 kg ha$^{-1}$ + one hand weeding at 5 WAE at Jari which was statistically similar to the weed free check at both locations (Table 7). Furthermore, the results showed that weed free check had also no

significant difference in number of pods plant$^{-1}$ obtained with the treatment of 1.0 kg ha$^{-1}$ of pendimethalin + one hand weeding at 5 WAE both at Jari and Sirinka as well as with the treatment of 1.0 kg ha$^{-1}$ of s-metolachlor + one hand weeding at 5 WAE at Sirinka. Two hand weedings at 2 and 5 WAE when interacted with the location did not show significant difference but proved significantly better than one hand weeding at 2 WAE at both locations.

The application of 1.0 kg ha$^{-1}$ of pendimethalin and 1.0 kg ha$^{-1}$ of s-metolachlor, each accompanied with one hand weeding resulted in significant increase in number of pods plant$^{-1}$ as compared to the application of these herbicides alone which was on account of prolonged weed control with hand weeding. This result is in line with the work of [6] who earlier stated that application of pendimethalin at 3.75 l ha$^{-1}$ + hand weeding at 5 weeks after sowing significantly gave higher mean values of yield components of cowpea. The more vigorous leaves under low infestation level helped to improve the photosynthetic efficiency of the crop and supported a large number of pods as reported from the work done earlier [33].

The lowest number of pods plant$^{-1}$ was observed in weedy check plots at Jari which was significantly lower than the other interactions except the interaction of 1.0 kgha$^{-1}$ and at 1.3 kg ha$^{-1}$ of pendimethalin at the same location. This might be due to the significantly more weed density and total weed dry weight (Table 3) in these treatments at Jari. The long season weed interference might have also resulted in shade effect that reduced the irradiance predominantly in the photosynthetically active region of the spectrum and the irradiance is a major ecological factor that influences plant growth [34].

These results are in line with [35] who observed an increased number of pods plant$^{-1}$ where weed population was reduced by management techniques. Similarly, [36,37] stated that the number of pods produced per plant or maintained to final harvest depends on a number of environmental and management practices.

#### 3.2.3.3. Number of seeds per pod

The number of seeds pod$^{-1}$ had a significant effect due to locations, and at Sirinka, the pods had 10.4% higher number of seeds compared to Jari. Despite a difference of 2.7 seeds pod$^{-1}$ the weed management practices did not show any significant difference (Table 5). However, in agreement with the findings of [30], this study also indicated lowest number of seeds pod[-1]in weedy check.

### 3.2.3.4. Hundred grain weight

The effect of locations and treatments was highly significant (P< 0.01) while their interaction had no significant effect on 100 seed weight. The seeds at Sirinka had significantly higher weight (by 7.6%) than at Jari. The grains under complete weed free plots recorded the highest weight which was statistically at par with two hand weeding at 2 and 5WAE, pendimethalin at 1.0 kg ha[-1] + one hand weeding at 5 WAE, s-metolachlor at 1.0 kg ha[-1]+ hand weeding at 5 WAE, pendimethalin at 1.3 kg ha[-1] and s-metolachlor 2.0 kg ha[-1]. The lowest 100 grain weight was observed in weedy check, however, it was comparable with one hand weeding at 2 WAE and s-metolachlor 1.5 kg ha[-1] (Table 6). The plants raised under complete weed free environment were free from weed competition. Thus, they utilized available resources to their maximum benefit leading to increased seed weight. Also, the more and vigorous leaves under weed free environment that improved the supply of assimilate to be stored in the seed, hence, the weight of 100 grains increased. The lowest 100 grain weight was observed in weedy check. However, it was equivalent with one hand weeding at 2 WAE and 1.5 kg ha[-1]of s-metolachlor (Table 6). This is consistent with [38] who stated that cowpea plants in unwedded plots gave the lowest 100 seed weight. However, [39 and 40] reported that there was no significant difference found in grain weight due to weed management practices in common bean.

### 3.2.3.5 Grain yield

Cowpea grain yield was significantly (P<0.01) influenced by the location, weed management practices and their interaction. The maximum grain yield (4277 kg ha[-1]) was obtained in complete weed free at Sirinka which was statistically at par with complete weed free at Jari and two hand weeding at Sirinka. Further, the interaction effect showed that the yield obtained with complete weed free treatment at Jari and two hand weeding at Sirinka had no significant difference with the application of1.0 kg ha[-1] of s-metolachlor + hand weeding at 5 WAE and 1.0 kg ha[-1] of pendimethalin + hand weeding at 5 WAE at Sirinka. It was also found that with the increasing rate of s-metolachlor application there

was an increase in yield but no significant variation was observed between 1.5 and 2.0 kg ha[-1] of s-metolachlor application at both locations (Table 7). But these treatments depicted significant yield increase over 1.0 kg ha[-1] of s-metolachlor application which was 22.5% and 33.0%, 18.7% and 22.7% over 1.0 kg ha[-1] of s-metolachlor, respectively at Jari and Sirinka.

The interaction effect of location with increasing rates of pendimethalin showed significant reduction in yield with the application of 1.3 and 1.6 kg ha[-1] of pendimethalin over its lower rate at Jari. In contrast, at Sirinka, results were inconsistent and no significant difference existed among the rates. At Jari, 1.0 kg ha[-1]of s-metolachlor + hand weeding at 5 WAE gave significant yield increase over 1.0 kg ha[-1] of pendimethalin + hand weeding at 5 WAE, while these weed management practices were statistically in parity at Sirinka. However, at both the locations, two hand weeding proved significantly better than one hand weeding (Table 7).

Weedy crop throughout the growing period resulted in the lowest grain yield, but at respective locations did not have a significant difference with 1.3 and 1.6 kg ha[-1] of pendimethalin at Jari, and 1.0 and 1.6 kg ha[-1] of pendimethalin as well as 1.0 kg ha[-1] of s-metolachlor at Sirinka. While comparing weedy check at Sirinka with weed management practices at Jari, the data (Table 5) revealed that weedy check plots had significantly higher yield than 1.0 kg ha[-1] of s- metolachlor, 1.3 and 1.6 kg ha[-1] of pendimethalin while it was statistically equivalent with 1.5 and 2.0 kg ha[-1] of s-metolachlor, 1.0 kg ha[-1] of pendimethalin and one hand weeding at 2 WAE (Table 7). The yield obtained at Sirinka in general, was significantly higher than at Jari under most of their respective weed management practices. This difference might have been partially due the differences that existed in number of pods plant [-1]and seeds pod[-1]between the locations. In line with this, [41] obtained significant increase in yield with the application of pendimethalin at 0.75 kg ha[-1] supplemented with one hand weeding 45 days after sowing in black gram. Similar conclusion has also been drawn by [42] that proper weed management gave higher yields of crops. The phenomenon involved in crop yield increase as affected by different weed control method has already been well described by [6,43,44,38] also stated that the yield and yield components of cowpea were also affected by weed control methods. This confirms the adverse effects of the weeds on the cowpea crop production sites as reported earlier by [45,46].

### 3.2.3.6. Aboveground dry biomass yield

The highest aboveground dry biomass yield (10797 kg ha$^{-1}$) was obtained in 1 kg ha$^{-1}$of s-metolachlor + one hand weeding at 5 WAE treated plots at Jari which was statistically at par with two hand weeding s at 2 and 5 WAE at the same location (9831 kg ha$^{-1}$), s-metolachlor at all the application rates (9815 to 10694 kg ha$^{-1}$), pendimethalin at 1.5 kg ha$^{-1}$, low rates of s-metolachlor and pendimethalin combined with one hand weeding at 5 WAE, one hand weeding and weed free check at Sirinka (Table 7). Weedy check plots had the lowest aboveground dry biomass yield among the treatments at respective locations, which was statistically at par with pendimethalin at 1.3 and 1.6 kg ha$^{-1}$ at Jari. At Sirinka, the aboveground dry biomass yield in weedy check was significantly lower than s-metolachlor at 2.0 kg ha$^{-1}$ and two hand weeding s at 2 and 5 WAE only. [47] reported that the increased dry matter weight of the crop was highly governed by the length of weed free period. While comparing the individual treatments in general, the aboveground dry biomass yield was higher at Sirinka than Jari. However, high production of total dry matter might not necessarily be of great value when the grain comprises a part of the plant. Though, the higher aboveground dry biomass in complete weed free and hand weeded plots may be due to better condition in soil rhizosphere that improved the competitive ability of the crop and favored more vegetative growth.

### 3.2.3.7. Harvest index

The results of both treatment revealed that the location and weed management practices had a significant influence on crop harvest index. The crop grown at Sirinka had significantly higher harvest index than at Jari. Highly significant harvest index was observed as compared to the weed management practices found when the crop was kept weed free throughout the growing season (Table 7). This was followed by pendimethalin 1.0 kg ha$^{-1}$ + hand weeding at 5 WAE, s-metolachlor 1.0 kg ha$^{-1}$+ hand weeding at 5 WAE and two hand weeding at 2 and 5 WAE. Though these weed management practices did not show statistically significant differences among them but had significantly higher harvest index than the remaining weed

management practices. The weedy check showed the minimum harvest index that did not significantly vary with 1.0 kg ha$^{-1}$ of s-metolachlor, 1.3 kg ha$^{-1}$ of pendimethalin and 1.6 kg ha$^{-1}$ of pendimethalin treatments. This lower harvest index might be due to severe weed competition with the crop for the growth factors, which restricted the growth and development of the crop in weedy check plots. Further, severe weed interference (Tables 2 and 3) might have decreased root/shoot ratio [46], increased vegetative growth duration (Table 4) and allocation of more assimilates for shoot rather than root growth. Likewise, the photosynthetic activity might be more during the vegetative phase of crop growth that contributed towards more total dry matter production, but the pace of this photosynthetic rate might have registered much higher decline due to disintegration of nodules with the initiation of pod development resulting in lower harvest index.

### 3.2.3.8. Yield loss

The weeds under different weed management practices caused variability in the amount of grain yield loss in cowpea. The highest yield loss (70.8%) was recorded in weedy check at Jari. This was statistically in parity with the loss registered with the application of 1.3 kg ha$^{-1}$ of pendimethalin and 1.6 kg ha$^{-1}$ of pendimethalin at the same location. All these weed management practices recorded a significant yield loss compared to other treatments. At Sirinka, the highest loss (47.5%) in yield, due to weeds, was also in weedy check, but it did not show significant variation with the loss accrued from the application of s- metolachlor at all rate, 1.0 kg ha$^{-1}$ pendimethalin and one hand weeding at 2 WAE at Jari, and 1.0 kg ha$^{-1}$ of s-metolachlor and 1.0 and 1.6 kg ha$^{-1}$of pendimethalin. The application of 1.0 kg ha$^{-1}$ of s-metolachlor combined with one hand weeding 5WAE resulted in the lowest yield loss which was statistically similar with two hand weeding at 2 and 5 WAE at both experimental sites, as well as 1.0 kg ha$^{-1}$ of s- metolachlor and 1.0 kg ha$^{-1}$ of pendimethalin were both combined with one hand weeding 5WAE at Sirinka. Moreover, it was observed that the yield loss due to s-metolachlor 1.0 kg ha$^{-1}$ superimposed with one hand weeding 5WAE at Jari and two hand weeding at 2 and 5 WAE at Sirinka were not significant as compared to complete weed free at both locations (Table 7). This observation is consistent with the work of [6,8] who reported that the presence of weeds reduced yield (by 82%).

**Table 4. Effect of weed management practices in cowpea on weed density (m⁻²) at harvest and on weed dry biomass (g m⁻²) at different growth stages of crop at Jari and Sirinka in 2013 main cropping season**

| Weed management practices | Weed density at harvest (m⁻²) | | | | | | Weed dry biomass weight (g m⁻²) | | | | | |
|---|---|---|---|---|---|---|---|---|---|---|---|---|
| | Broadleaved | | Sedges | | Total | | 20 DAE | | 55 DAE | | At harvest | |
| | Jari | Sirinka | Jari | Sirinka | Jari | Sirinka | Jari | Sirinka | Jari | Sirinka | Jari | Sirinka |
| S-metolachlor at 1.0 kg ha⁻¹ | 107.67$^d$ | 38.67$^c$ | 29.00$^d$ | 4.33$^d$ | 136.7$^d$ | 43.00$^c$ | 75.4$^f$ | 15.77$^d$ | 295.9$^e$ | 124.27$^c$ | 312.3$^d$ | 138.7$^b$ |
| S-metolachlor at1.5 kg ha⁻¹ | 61.67$^e$ | 20.67$^e$ | 11.33$^e$ | 3.00$^{de}$ | 73.0$^e$ | 23.67$^e$ | 48.8$^g$ | 12.73$^e$ | 183.4$^f$ | 92.40$^e$ | 209.9$^e$ | 98.7$^c$ |
| S-metolachlor at 2.0 kg ha⁻¹ | 12.00$^f$ | 3.00$^{fg}$ | 1.33$^f$ | 1.67$^{de}$ | 13.3$^f$ | 4.67$^{fg}$ | 22.0$^h$ | 10.27$^f$ | 124.2$^h$ | 82.30$^f$ | 146.9$^g$ | 102.0$^c$ |
| Pendimethalin at1.0 kg ha⁻¹ | 152.00$^c$ | 29.33$^d$ | 40.67$^c$ | 4.33$^d$ | 192.7$^c$ | 33.67$^d$ | 108.5$^e$ | 14.47$^{de}$ | 504.9$^d$ | 116.33$^d$ | 541.0$^c$ | 130.2$^{bc}$ |
| Pendimethalin at1.3 kg ha⁻¹ | 176.00$^b$ | 52.00$^b$ | 55.33$^b$ | 13.67$^c$ | 231.3$^b$ | 65.67$^b$ | 150.2$^c$ | 19.77$^c$ | 547.3$^c$ | 128.07$^{bc}$ | 574.4$^b$ | 142.1$^b$ |
| Pendimethalin at 1.6 kg ha⁻¹ | 146.67$^c$ | 52.00$^b$ | 39.33$^c$ | 17.67$^b$ | 186.0$^c$ | 69.67$^b$ | 133.3$^d$ | 21.67$^c$ | 559.8$^b$ | 128.60$^b$ | 581.5$^b$ | 144.8$^b$ |
| S-metolachlor at 1.0 kg ha⁻¹+hand weeding at 5 WAE | 1.33$^f$ | 2.67$^{fg}$ | 1.33$^f$ | 3.00$^{de}$ | 2.7$^f$ | 5.67$^f$ | 71.8$^f$ | 14.40$^{de}$ | 27.3$^j$ | 21.50$^h$ | 43.5$^f$ | 29.0$^d$ |
| Pendimethalin at 1.0 kg ha⁻¹+hand weeding at 5 WAE | 1.33$^f$ | 5.00$^f$ | 0.00$^f$ | 3.00$^{de}$ | 1.3$^f$ | 8.00$^f$ | 105.4$^e$ | 12.23$^{ef}$ | 27.4$^j$ | 25.70$^g$ | 38.1$^f$ | 32.3$^d$ |
| One hand weeding at 2 WAE | 54.67$^e$ | 2.67$^{fg}$ | 12.67$^e$ | 2.33$^{de}$ | 67.3$^e$ | 5.00$^{fg}$ | 172.6$^b$ | 21.56$^c$ | 142.4$^g$ | 116.90d | 170.4$^f$ | 126.8$^{bc}$ |
| Hand weeding at 2 and 5WAE | 5.33$^f$ | 3.67$^{fg}$ | 2.67$^f$ | 2.67$^{de}$ | 8.0$^f$ | 6.33$^f$ | 173.6$^b$ | 25.90$^b$ | 96.5$^i$ | 82.73$^f$ | 110.0$^h$ | 98.0$^c$ |
| Weed free check | 0.00$^f$ | 0.00$^g$ | 0.00$^f$ | 0.00$^e$ | 0.0$^f$ | 0.00$^g$ | 0.0$^i$ | 0.00$^g$ | 0.0$^k$ | 0.00$^i$ | 0.0$^j$ | 0.0$^d$ |
| Weedy check | 289.33$^a$ | 92.00$^a$ | 80.67$^a$ | 24.33$^a$ | 370.0$^a$ | 116.33$^a$ | 178.8$^a$ | 32.20$^a$ | 882.8$^a$ | 265.40$^a$ | 906.3$^a$ | 247.6$^a$ |
| LSD (5%) | 16.12 | 4.64 | 7.84 | 3.37 | 20.28 | 5.42 | 4.38 | 2.43 | 6.30 | 3.96 | 16.36 | 32.30 |
| CV (%) | 11.3 | 10.9 | 20.3 | 29.9 | 11.2 | 10.1 | 2.5 | 8.6 | 1.3 | 2.4 | 3.2 | 17.7 |

CV= coefficient of variation, LSD= least significant difference, DAE= days after emergence, WAE= weeks after emergence, Means in coloumns of same parameter followed by the same letter(s) are not significantly different at 5% level of significance

**Table 5. Interaction effect of location and weed management practices on days to flowering and physiological maturity of cowpea in 2013 main cropping system**

| Weed management practices | Days to 50% flowering | | Days to physiological maturity | |
|---|---|---|---|---|
| | Jari | Sirinka | Jari | Sirinka |
| S-metolachlor at 1.0 kg ha[-1] | 61.0[d] | 62.0[c] | 85.0[c] | 94.0[a] |
| S-metolachlor at 1.5 kg ha[-1] | 61.0[d] | 62.0[c] | 85.0[c] | 94.0[a] |
| S-metolachlor at 2.0 kg ha[-1] | 59.0[e] | 57.3[f] | 83.0[e] | 92.0[b] |
| Pendimethalin at 1.0 kg ha[-1] | 61.0[d] | 62.0[c] | 85.0[c] | 94.0[a] |
| Pendimethalin at 1.3 kg ha[-1] | 61.0[d] | 62.0[c] | 85.0[c] | 94.0[a] |
| Pendimethalin at 1.6 kg ha[-1] | 61.0[d] | 62.0[c] | 85.0[c] | 94.0[a] |
| S-metolachlor at 1.0 kg ha[-1]+ hand weeding at 5 WAE | 56.7[g] | 55.0[h] | 83.0[e] | 92.0[b] |
| Pendimethalin at 1.0 kg ha[-1] + hand weeding at 5 WAE | 57.7[f] | 56.3[g] | 83.0[e] | 92.0[b] |
| One hand weeding at 2 WAE | 61.0[d] | 62.0[c] | 85.0[c] | 94.0[a] |
| Two hand weeding at 2 and 5 WAE | 61.0[d] | 62.0[c] | 83.7[d] | 94.0[a] |
| Weed free check | 57.3[f] | 55.7[h] | 83.0[e] | 92.0[b] |
| Weedy check | 63.0[b] | 64.0[a] | 85.0[c] | 94.0[a] |
| LSD (5%) L x WMP | 0.63 | | 8.0 | |
| CV (%) | 0.6 | | 0.3 | |

*CV= coefficient of variation, LSD= least significant difference, WAE= weeks after emergence, Means in coloumn and row of same parameter followed by the same letter(s) are not significantly different at 5% level of significance*

**Table 6. Effect of location and weed management practices on plant height (cm) of cowpea in 2013 main cropping season**

| Treatments | Plant height (cm) |
|---|---|
| **Location** | |
| Jari | 68.07[a] |
| Sirinka | 60.60[b] |
| LSD (5%) | 3.82 |
| **Weed management practices** | |
| S-metolachlor at 1.0 kg ha[-1] | 64.04 |
| S-metolachlor at1.5 kg ha[-1] | 65.53 |
| S-metolachlor at 2.0 kg ha[-1] | 61.49 |
| Pendimethalin at1.0 kg ha[-1] | 66.52 |
| Pendimethalin at1.3 kg ha[-1] | 68.20 |
| Pendimethalin at 1.6 kg ha[-1] | 62.51 |
| S-metolachlor at 1.0 kg ha[-1]+ hand weeding at 5 WAE | 62.87 |
| Pendimethalin at 1.0 kg ha[-1]+ hand weeding at 5 WAE | 61.47 |
| One hand weeding at 2 WAE | 64.56 |
| Two hand weeding at 2 and 5 WAE | 60.06 |
| Weed free check | 62.26 |
| Weedy check | 73.58 |
| LSD (5%) | NS |
| CV (%) | 12.51 |

*CV= coefficient of variation, LSD= least significant difference, WAE= weeks after emergence, NS= not significant, Means in coloumn followed by the same letter(s) are not significantly different at 5% level of significance*

On the other hand, [7,47,48] found that there existed 12.7%-60.0%,40% to 60% and 25% and 60% reduction in potential yield of cowpea due to weeds, respectively. This difference in reduction in cowpea yield reported by various researchers might be due to the differences in weed flora, crop varieties and environmental conditions prevailing in the study area. Therefore, the difference in time of weed removal might have contributed to this variation in yield. The herbicide might have dissipated soon from the soil under the influence of environmental conditions prevailing during the crop season.

## 3.3 Partial Budget Analysis

The result of the partial budget analyses showed that two hand weeding accrued 4.2 and 13.4% higher total variable cost than 1.0 kg ha[-1] of pendimethalin and1.0 kg ha[-1] of s–metolachlor both superimposed with hand weeding, respectively at both sites (Table 9). On the other hand highest gross as well as net benefits were obtained with the application of 1.0 kg ha[-1]of s-metolachlor at + hand weeding , followed by two hand weeding at 2 and 5 WAE and 1.0 kg ha[-1]of pendimethalin +hand weeding.

**Table 7. Effect of weed management practices on crop stand (ha⁻¹), number of seeds pod⁻¹, hundred seed weight (g) and harvest index (%) of cowpea at Jari and Sirinka in 2013 main cropping season**

| Factors | Crop stand (ha⁻¹) | Number of seeds pod⁻¹ | Hundred grain weight (g) | Harvest index |
|---|---|---|---|---|
| **Location** | | | | |
| Jari | 98251[b] | 12.01[b] | 11.90[b] | 0.28[de] |
| Sirinka | 106224[a] | 13.26[a] | 12.82[a] | 0.31[a] |
| LSD (5%) | 4830.1 | 0.78 | 0.27 | 0.013 |
| CV (%) | 10.0 | 13.05 | 4.70 | 9.10 |
| **Weed management practices** | | | | |
| S-metolachlor at 1.0 kg ha⁻¹ | 92978[d] | 12.03 | 12.39[bc] | 0.24[de] |
| S-metolachlor at 1.5 kg ha⁻¹ | 96065[cd] | 13.08 | 11.80[cd] | 0.30[c] |
| S-metolachlor at 2.0 kg ha⁻¹ | 105324[bc] | 12.74 | 12.64[ab] | 0.29[c] |
| Pendimethalin at 1.0 kg ha⁻¹ | 95293[cd] | 12.85 | 12.24[bc] | 0.27[cd] |
| Pendimethalin at 1.3 kg ha⁻¹ | 93750[cd] | 12.56 | 12.66[ab] | 0.24[de] |
| Pendimethalin at 1.6 kg ha⁻¹ | 101466[cd] | 11.36 | 12.32[bc] | 0.25[de] |
| S-metolachlor at 1.0 kg ha⁻¹ + hand weeding  at 5 WAE | 120756 [a] | 13.40 | 12.68[ab] | 0.35[b] |
| Pendimethalin at 1.0 kg ha⁻¹ + hand weeding  at 5 WAE | 117284 [a] | 13.60 | 12.55[ab] | 0.37[b] |
| One hand weeding  at 2 WAE | 90664 [d] | 12.72 | 12.05[bcd] | 0.29[c] |
| Two hand weeding  at 2 and 5 WAE | 114969[ab] | 12.57 | 12.52[ab] | 0.36[b] |
| Weed free check | 120370 [a] | 13.68 | 13.11[a] | 0.41[a] |
| Weedy check | 77932 [e] | 11.02 | 11.44[d] | 0.22[e] |
| LSD (5%) | 11831.3 | NS | 0.67 | 0.032 |
| CV (%) | 10.0 | 13.05 | 4.7 | 9.10 |

CV= coefficientof variation; LSD= least significant difference; WAE= weeks after emergence; NS= not significant; Means in coloumn of same parameter followed by the same letter(s) are not significantly different at 5% level of significance

**Table 8. Interaction effect of location and weed management practices on number of pods plant⁻¹, grain and aboveground dry biomass yield (kg ha⁻¹) and yield loss (%) in cowpea in 2013 main cropping season**

| Weed management practices | Number of pods plant⁻¹ | | Grain yield (kg ha⁻¹) | | Aboveground dry biomass yield (kg ha⁻¹) | | Yield loss (%) | |
|---|---|---|---|---|---|---|---|---|
| | Jari | Sirinka | Jari | Sirinka | Jari | Sirinka | Jari | Sirinka |
| S-metolachlor at 1.0 kg ha⁻¹ | 12.11[hij] | 12.70[hij] | 1750[l] | 2595[hij] | 7408[fg] | 10082[abc] | 55.4[b] | 39.4[cde] |
| S-metolachlor at1.5 kg ha⁻¹ | 16.00[def] | 11.50[ij] | 2144[kl] | 3080[efg] | 7185[gh] | 9815[abcd] | 44.7[cd] | 28.0[fg] |
| S-metolachlor at 2.0 kg ha⁻¹ | 13.11[ghi] | 17.13[cde] | 2327[ijk] | 3185[def] | 7780[fg] | 10694[ab] | 39.9[cde] | 25.6[gh] |
| Pendimethalin at1.0 kg ha⁻¹ | 10.67[ijk] | 14.53[tgh] | 2373[ijk] | 2582[hij] | 8957[de] | 9136[cde] | 39.2[cde] | 39.6[cde] |
| Pendimethalin at1.3 kg ha⁻¹ | 8.00[l] | 12.83[hij] | 1322[m] | 2696[ghi] | 6157[hi] | 9763[abcd] | 66.1[a] | 36.9[def] |
| Pendimethalin at 1.6 kg ha⁻¹ | 8.67[kl] | 15.33[efg] | 1282[m] | 2555[hijk] | 5932[i] | 9059[cde] | 67.1[a] | 40.1[cde] |
| S-metolachlor at 1.0 kg ha⁻¹ + hand weeding at 5 WAE | 22.44[a] | 19.07[bc] | 3595[cd] | 3769[bc] | 10797[a] | 9949[abcd] | 7.9[ij] | 11.8[i] |
| Pendimethalin at 1.0 kg ha⁻¹ + hand weeding at 5 WAE | 19.33[bc] | 18.97[bc] | 3017[fg] | 3614[bc] | 8161[efg] | 9753[abcd] | 22.7[gh] | 15.3[hi] |
| One hand weeding at 2 WAE | 15.33[efg] | 15.83[def] | 2312[ijk] | 2969[fgh] | 8382[ef] | 9733[abcd] | 40.1[cde] | 30.5[etg] |
| Two hand weeding at 2 and 5 WAE | 18.11[cd] | 19.00[bc] | 3452[de] | 3864[abc] | 9831[abcd] | 10438[ab] | 11.6[i] | 9.5[ij] |
| Weed free check | 20.78[ab] | 20.67[ab] | 3907[ab] | 4277[a] | 9566[bcd] | 10113[abc] | 0.0[j] | 0.0[j] |
| Weedy check | 7.67[l] | 10.63[ijk] | 1134[m] | 2241[jk] | 5661[i] | 9043[cde] | 70.9[a] | 47.5[bc] |
| LSD (%) L x WMP | 2.456 | | 422.0 | | 1079.0 | | 10.57 | |
| CV(%) | 10.0 | | 9.3 | | 7.4 | | 19.6 | |

CV= coefficient of variation; LSD= least significant difference; WAE= weeks after emergence; Means in coloumn and row of same parameter followed by the same letter(s) are not significantly different at 5% level of significance

**Table 9. Partial budget analysis of weed management practices in cowpea based on total variable cost in 2013 main cropping season**

| Weed management practices | Total variable cost (ETB ha$^{-1}$) | Average yield (kg ha$^{-1}$) | Adjusted yield (kg ha$^{-1}$) 10% down | Gross benefit (ETB ha$^{-1}$) | Net Benefit (ha$^{-1}$) |
|---|---|---|---|---|---|
| S-metolachlor at 1.0 kg ha$^{-1}$ | 3935 | 2172.4 | 1955.2 | 29328 | 25393 |
| S-metolachlor at 1.5 kg ha$^{-1}$ | 4841 | 2612.1 | 2350.9 | 35264 | 30423 |
| S-metolachlor at 2.0 kg ha$^{-1}$ | 5283 | 2756.0 | 2480.4 | 37206 | 31923 |
| Pendimethalin at 1.0 kg ha$^{-1}$ | 5589 | 2477.5 | 2229.8 | 33447 | 27858 |
| Pendimethalin at 1.3 kg ha$^{-1}$ | 5339 | 2009.0 | 1808.1 | 27122 | 21783 |
| Pendimethalin at 1.6 kg ha$^{-1}$ | 5581 | 1918.6 | 1727.4 | 25911 | 20330 |
| S-metolachlor at 1.0 kg ha$^{-1}$ +hand weeding at 5 WAE | 6828 | 3682.0 | 3313.8 | 49707 | 42879 |
| Pendimethalin at 1.0 kg ha$^{-1}$ +hand weeding at 5 WAE | 7430 | 3315.8 | 2984.2 | 44763 | 37333 |
| One hand weeding at 2 WAE | 5620 | 2640.5 | 2376.5 | 35648 | 35649 |
| Two hand weeding at 2 and 5 WAE | 7742 | 3658.4 | 3292.6 | 49380 | 41638 |
| Weedy check | 2642 | 1687.3 | 1518.6 | 22779 | 20137 |

*Cost of s-metolachlor 417 Birr/ kg; cost of pendimethalin 620 Birr/kg; Spraying Birr 99/ ha; Cost of hand weeding and hoeing 2 WAE 45 persons, 35 DAE 16 persons @Birr 33 / person; Sale price of cowpea Birr 15/ kg; Field price of cowpea (sale price- variable input cost-harvesting, threshing and winnowing Birr 165/ 100 kg; packing and material cost Birr 4.0 per 100 kg, transportation Birr 5 per 100 kg ; ETB= 0.0498 USD*

## 4. CONCLUSION

It can be inferred from this study that the treatment of 1.0 kg ha$^{-1}$ of s-metolachlor supplemented with one hand weeding at 5 DAE is the most profitable treatment with an alternate weed management option i.e. 1.0 kg ha$^{-1}$ of pendimethalin supplemented with one hand weeding 35 DAE. Therefore, managing the weeds with the application of 1.0 kg ha$^{-1}$ of s-metolachlor + hand weeding and hoeing at 35 DAE proved to be the most profitable practice. However, under the condition of labour constraint and timely availability of the herbicide, pre emergence application 2.0 kg ha$^{-1}$ of s-metolachlor should be used to preclude the yield loss and ensure maximum benefits. For broad-spectrum and effective weed control, the herbicide mixture should also be tested. Further, to prevent the weed shift, these two herbicides should be used as herbicide rotation. In future, there is a need to explore the effectiveness of various combinations of these two herbicides for cost effective and broad spectrum weed control in cowpea production.

## ACKNOWLEDGMENTS

Above all, I wish to thank Jesus Christ, my good Shepherd, Savior and Lord, "for everything that was made through Him, nothing was made without Him." John 1:3. I would like to express my passionate thankfulness to my beloved wife Mekdes Dessie for her moral and sensible support and inspiration during the course of my study. I wish to express my deep, heartfelt gratitude to my Mother Tureya Musa for her help, pray and motivation during my under and post graduate study that paved the way for today's success. Last but not least, I am grateful to Bezawit Mewa for funding my paper expenses.

## COMPETING INTERESTS

Authors have declared that no competing interests exist.

## REFERENCES

1.    FAO. World Agriculture: towards 2015/2030. Summary report, Rome. 2002;45:3,492–515.

2.    Agbogidi O. M. Screening six cultivars of cowpea [*Vigna unguiculata* (L.) Walp.] for adaptation to soil contaminated with spent engine oil. Journal of Environmental Chemistry and Ecotoxicology. 2010;7:103-109.

3.    Muoneke CO, Ndukwe OM, Umana PE, Okpara DA, Asawalam DO. Productivity of vegetables cowpea [*Vigna unguiculata* (L.) Walp.] and maize (*Zea mays* L.) intercropping system as Influenced by component density in a tropical zone of

southeastern Nigeria. International Journal of Agricultural Research and Development. 2012;15:835-847.

4.   Shiringani RP, Shimeles HA. Yield response and stability among cowpea genotypes at three planting dates and test environments. African Journal of Agricultural Resources. 2011;6:3259-3263.

5.   Sankie L, Addo-Bediako KO, Ayodele V. Susceptibility of seven cowpea [*Vigna unguiculata* (L.) Walp.]Cultivars to cowpea beetle *(Callosbruchus maculatres)*. Agricultural Science Research Journals. *2012;2:65-69.*

6.   Patel MM, Patel AI, Patel IC, Tikka SBS, Henry A, Kumar D, Singh NB. Weed control in cowpea under rain fed conditions. In: Proceedings of the National Symposium on Arid Legumes, for Food Nutrition. Security and promotion of Trade. Advances in Arid Legumes Research. Hisar, India. 2003;203-206.

7.   Li RG, Yumei Z, Zhanzhi X. Damage loss and control technology of weeds in cowpea field. Journal of Weed Science. 2004;2:25-26.

8.   Muhammad RC, Muhammad J, Tahira ZM. Yield and yield components of cowpea as affected by various weed control methods under rain fed conditions of Pakistan. International Journal of Agriculture and Biology. 2003;09:120-124.

9.   Khan IG, Hassan MI, Khan MI, Khan IA. Efficacy of some new herbicidal molecules on grassy and broadleaf weeds in wheat-II. Pakistan Journal of Weed Science Research. 2004;10:33-38.

10.  Akobundu A. Weed Science in the Tropics. Principles and Practices. 2nd ed, John Wiley and Sons Inc. Chicheste; 1987.

11.  Dadari SA. Evaluation of herbicides in cowpea or cotton mixture in Northern Guinea Savannah. Journal of Sustainable Agriculture and Environment. 2003;5;153-159.

12.  Jaibir T, Singh T, Vivek HB, Tripathi SS. Integrated weed management in intercropping of mungbean (*Vigna radiata*) and cowpea fodder (*Vigna unguiculata*) with pigeonpea (*Cajanus cajan*) under western U.P. condition. Indian Journal of Weed Science. 2004;36:133-134.

13.  Lagoke STO, Choudhary AH, Chandra-Singh DJ. Chemical Weed Control in Rainfed Cowpea [*Vigna unguiculata* (L.) Walp.] in the Guinea Savanna Zone of Nigeria. Weed Research. 1982;22:17-22.

14.  Stroud A. Weed Management in Ethiopia. An Extension and Training Manual. FAO, Rome. 1989;4:190-193

15.  Payne RW, Murray DA, Harding SA, Baird DB, Soutar DM. Gen Stat for windows (12nd edn.) Introduction. VSN International, Hemel, Hempstead; 2009.

16.  Gomez KA, Gomez AA. Statistical procedures for Agricultural Research 2d edition. John Wiley and Sons Inc., New York. 1984;145.

17.  CIMMYT, from agronomic data to farmer recommendations: Answers to workbook exercises. Mexico; 1988.

18.  Khan MA, Marwat KB, Khan N, Khan IA. Efficacy of different herbicides on the yield and yield components of maize. Asian Journal of Plant Science. 2003;2:300-304.

19.  Holt JS. Plant response to light: A potential tool for weed management. Weed Science. 1995;43:474-482.

20.  Hooda IS, Agrawal SK. Response of weeds to levels of irrigation, weed control and fertility in wheat. British Crop Protection Conference: Weeds. In: Proceedings of International Conference, Brighton, U.K. 1995;2:679-682.

21.  Hafeezzullah. Effect of different sowing and weed control methods on the performance of sunflower. M. Sc (Hons) Thesis, Department of Agronomy, Agricultural University, Peshawar, Pakistan; 2000.

22.  Mahale SS. Integrated weed control measures on weed growth, yield and yield attributes in rainfed groundnut. Indian Journal of Agricultural Sciences. 1992;65: 42-45.

23.  Gonzalez PR, Salas ML. Weed control with metolachlor and atrazine in maize: Effects on yield and nutrition of the crop. In: Proceedings of Cong Spanish Weed Science Society, Huesca, Spain. 1995;193-198.

24.  Sharma GD, Sharma JJ, Sood S. Evaluation of alachlor, metolachlor and pendimethalin for weed control in rajmash (*Phaseolus vulgaris* L.) in cold desert of North-Western Himalayas. Indian Journal of Weed Science. 2004;36:287-289.

25.  Sharma V, Thakur DR, Sharma JJ. Effect of metolachlor and its combination with atrazine on weed control in maize (*Zea mays*). Indian Journal of Agronomy. 1998;43:677-680.

26. Kumar S, Sharma GD, Sharma JJ. Integrated Weed Management studies in Rajmash (*Phaseolus vulgaris*) under dry temperate high-hills. Indian Journal of Weed Science. 1997;28:8-10.

27. Rana SS. Evaluation of Promising herbicide Combinations for weed management in Rajmash under dry temperate condition of Himachal Pradesh. Indian Journal of Weed Science. 2002;34:204-207.

28. Kumar NS. Effect of plant density and weed management practices on production potential of groundnut (*Arachis hypogaea* L.). Indian Journal of Agricultural Research, 2009;43:57-60.

29. Das TK, Yaduraju NT. Effect of weed competition on growth, nutrient uptake and yield of wheat as affected by irrigation and fertilizers. Journal of Agricultural Science, 1999;133(1):45-51

30. Gupta OP. The nature of weed competition. In: Modern Weed Management with special reference to agriculture in the tropics and sub tropics. A text book and manual. Agrobios, India. 2011;634.

31. Sunday Omovbude and Ekea Udensi. Evaluation of Pre-Emergence Herbicides for Weed Control in Cowpea [*Vigna unguiculata* (L.) Walp.] in a Forest - Savanna Transition Zone. American Journal of Experimental Agriculture. 2013;3: 767-779.

32. Chattha MR, Jamit M, Mahmood TZ. Yield and yield components of cowpea as affected by various control methods and rain-fed condition of Pakistan. International Journal of Agricultural Biology. 2007;9(1):120-124.

33. Kamel MS, Abdel-Raouf MS, Mahmoud EA, Amer S. Response of two maize varieties to different plant densities in relation to weed control treatments. Annals of Agricultural Science. 1983;19:79-93.

34. Abdellatif YI. Effect of seed size and plant spacing on yield and yield components of faba bean (*Vicia faba* L.). Research Journal of Agriculture and Biological Sciences. 2008;4(2):146-148.

35. Hadi H, Ghassemi-Golezani K, Khoei FR, Valizadeh M, Shakiba MR. Response of common bean (*Phaseolus vulgaris* L.) to different levels of shade. Journal of Agronomy. 2006;5:595-599.

36. Yadav RP, Shrivastava UK, Dwivedi SC. Comparative efficiency of herbicides in controlling *Asphodelus tenuifolius* and other weeds in Indian mustard (*Brassica juncea* L.). Indian Journal of Agronomy. 1999;44:151-155.

37. Ayaz S, McNeil DL, McKenzie BA, Hill GD. Density and sowing depth effects on yield components of grain legumes. In: Proceeding of Agronomy Society, New Zealand. 2001;13:81-86,9-15.

38. Madukwe DK, Ogbuehi HC, Onuh MO. Effects of Weed Control Methods on the Growth and Yield of Cowpea [*Vigna unguiculata*(L.) Walp.] under Rain-Fed Conditions of Owerri. American-Eurasian Journal Agriculture and Environmental Science.2012;11:1426-1430.

39. Meseret Negash, Tadese Berhanu and Teshome Bogale. Effect of frequency and time of hand weeding in common bean production at Bako. Ethiopian Journal of Weed Management. 2008;2:59-69.

40. Kumar S, Angiras NN, Singh R. Effect of planting and weed control methods on weed growth and seed yield of Black gram. Indian Journal of Weed Science. 2006;38:73-76.

41. Roslon E, Fogelfors H. Crop and weed growth in a sequence of spring barley and winter wheat crops established together from a spring sowing (relay cropping). Journal of Agronomy and Crop Sciences. 2003;189:185–190.

42. Bukhtiar BA, Naseem BA, Tufail M. Weed control in lentil under irrigated conditions. Pakistan Journal Weed Sciences Research. 1991;4:99–104.

43. Mathew G, Sreenivasan E. Effect of weed control methods on yield and economics of rain-fed and rice fallow summer cowpea. Madras Agriculture Journal, 1998;85:50–52.

44. Tomar RK, Singh JP, Garg RN, Gupta VK, Sahoo RN, Arora RP. Effect of weed management practices on weed growth and yield of wheat in rice based cropping system under varying levels of tillage. Annals Plant Protection Sciences. 2003;11:123–128.

45. Tijani EH. Influence of intra row spacing and weeding regime on the performance of cowpea [*Vigna unguiculata* (L.) Walp.]. Nigerian Journal of Weed Sciences. 2001;14:11-15.

46. Mizan A, Sharma JJ, Gebremedhin W. Estimation of critical period of weed-crop competition and yield loss in sesame

(*Sesamum indicum* L.). Ethiopian Journal of Weed Management. 2009;3(1):39-53.

47. Ishaya DB, Tunku P, Kuchinda NC. Evaluation of some weed control treatments for long season weed control in maize under zero and minimum tillage at Samaru in Nigeria. Crop Protection. 2008;27:1047-1051.

48. Joseph A, Osipitan AO, Segun TL, Raphael OA, Stephen OA. Growth and yield performance of cowpea [*Vigna unguiculata* (L.) Walp.] as influenced by row-spacing and period of weed interference in South-West Nigeria. Journal of Agricultural Science. 2014;4:1916-9760.

# 12

# Effects of Jatropha Seed Cake (JSC) and Different Inorganic Fertilizers on Growth and Yield of Two Groundnut Cultivars under Three Harvesting Periods

## S. M. H. Elseed[1], S. O. Yagoub[2*] and I. S. Mohamed[3]

[1]Ministry of Agriculture and Animal Resources and Irrigation, Khartoum State, Sudan.
[2]Department of Agronomy, College of Agricultural Studies, Sudan University of Science and Technology Box, 73, Sudan.
[3]Department of Plant Protection, College of Agricultural Studies, Sudan University of Science and Technology, Sudan.

*Authors' contributions*

This work was carried out in collaboration between all authors. This work was carried out in collaboration between three authors. Author SMHE initiated the experiments, collected the data, performed the statistical analysis, Author SOY designed the study, managed the literature review and wrote the first drafts of the manuscript. Author ISM assisted with statistical analysis and contributed to the final draft. Three authors read and approved the final manuscript.

*Editor(s):*
(1) Aleksander Lisowski, Warsaw University of Life Sciences, Department Agricultural and Forestry Engineering, Poland.
(2) Daniele De Wrachien, State University of Milan, Italy.
*Reviewers:*
(1) A.I. Gabasawa, Department of Soil Science, Ahmadu Bello University, Zaria, Nigeria.
(2) Anonymous, National Semi-Arid Research Resources Institute, Uganda.

## ABSTRACT

**Aim:** to investigate the effect of Jatropha Seed Cake (JSC) and different fertilizers on growth and yield of two groundnut cultivars under three harvesting periods
**Study Design:** The experimental design was a randomized complete block in a split-plot replicated three times. Groundnut cultivars arranged as whole plots and five fertilizers treatment as sub- plot.
**Locations:** field experiments were conducted for two seasons 2011/12-2012/13 under semi-arid condition of Khartoum state in demonstration farm of Sudan University of Science and Technology,

*Corresponding author: Email: umreelah2003@yahoo.com;*

Khartoum North, Shambat.

**Material and methods:** The treatments were Jatropha Seed Cake (2.5 t/ha), sulfur (119 kg/ha), single super phosphate (119 kg/ha), ammonium sulphate (119 kg/ha) and control on two groundnut cultivars (Sodri) and (Gebish) under three harvesting time (every 10 days).

**Results:** The results showed that yield increased with delayed in harvest periods for two seasons. The two tested cultivars revealed similar behavior during vegetative and reproductive growth. Except for plant height of third harvest of first season, primary branch of second harvest and yield of first and third harvest of two seasons. Results showed significant difference among fertilizers treatments and interaction between cultivars and fertilizers concerning the majority of growth and yield components. Super phosphate gave significant difference on leaf area index, plant height, number of primary branches, number of primary branches, number of pods/plant, germination %. On the other hand, sulfur and ammonium sulphate gave significant difference on number of pods/plant, germination % and yield. Ammonium sulphate and Jatropha Seed Cake reviled significant effect on yield in first and second season respectively

**Conclusion:** In conclusion addition of Jatropha Seed Cake and inorganic fertilizers appeared to be more promising. Delaying harvesting periods gave positive results.

Keywords: Groundnut inorganic fertilizers; Jatropha seed cake.

## 1. INTRODUCTION

Groundnut Arachis hypogaea L is a major cash crop grown mainly under rain-fed conditions in the semi-arid tropics. Groundnut is the 13th most important food crop of the world. It is the world's 4th most important source of edible oil and 3rd most important source of vegetable protein [1]. Groundnut is primarily used for oil extraction in Sudan [2]. It is consumed directly because of its high food value. Sudan is one of the major groundnut producing countries. The total area under groundnut production is approximately one million ha with an average yield of 855 kg/ha [2].

Substantial evidence showed that groundnut responds well to additional inorganic and organic fertilizers application [3-9]. Phosphorus is very important nutrient element for crop growth and yield. It plays an important role in physiological processes of plants. As P source, single superphosphate is the most suitable fertilizer for groundnut in Nigeria Savannah [10]. Sulfur deficiency in legume crops affects not only yield, but also the nutritional quality of the seeds [11].

Biodiesel is an environmental-friendly substitute for fossil fuel. Research on the production of biodiesel from vegetable oils has concentrated on jatropha, soybean, palm kernel, sunflower and rapeseed oils with scarce information on groundnut oils [12]. The biodiesel from Jatropha curcas had received much attention in past decades. The byproduct of oil extraction from Jatropha curcas seeds and kernel is called seedcake. It is containing highly toxic protein that is not suitable for animal's feeds, although good for organic manure; it is being used as an organic fertilizer [13]. The cake is rich in nitrogen (3.2%) phosphorus (1.4%) and potassium (1.3%) and can be used as manures [14 and 15]. Limited studies have been reported on potential of Jatropha Seed Cake to be used as organic fertilizers. Organic manures are very important, as it contains both major and minor elements necessary for plant growth, and improve the physical, chemical and biological properties of soil [16]. The productivity of groundnut depends on proper selection of variety, fertilizer management. This study was carried out to identify the effect of Jatropha Seed Cake (JSC) and different types fertilizers on growth and yield of two groundnut cultivars under three harvesting periods.

## 2. MATERIALS AND METHODS

### 2.1 Experimental site

Field experiments were conducted in Demonstration farm, College of Agricultural Studies, Sudan University of Science and Technology, Shambat, Khartoum North, Sudan for two consecutive seasons 2011/12- 2012/13. Shambat is located at longitude 32 35"E and latitude 15 31"N, within the semi-desert region. Climate of the locality is semi-desert and tropical with low relative humidity.

## 2.2 Layout of the Experiment and Land Preparation

The experiment was designed in split plots replicated three times. Two cultivars of groundnut (Sodri =V1) and (Gebish=V2) and four types of fertilizers were applied; (F1) control without fertilizer, (F2) inorganic Jatropha Seed Cake (JSC) of 2.5 t/ha, and three organic fertilizers: (F3) pure sulfur of 119 kg/ha, (F4) super phosphate of 119 kg/ha and (F5) Ammonia sulphate of 119 kg/ha. After maturing (about 110 days from sowing) three harvesting periods were taken (every 10 days), early, medium and late. The field was prepared according to adopted method by Engineering Department of Agricultural Colleage. The seeds were sown in July 12th, and July 21th respectively for two seasons, 3-4 seeds per hole then thinned to two seed per hole. During July until August (about two months) it was irrigated by rain-fed then irrigation was applied every 15 days. For first season the three harvesting time were, October, 11- 21 and first November, and for second season October, 24, and November 4-14.

## 2.3 The Source of Seeds

Two cultivars of groundnut certified seed (Sodri = V1) and (Gebish= V2) were obtained from Al – Obied Research Center (North Kordfan), and Arabian Company for Seed, Khartoum.

## 2.4 Data Collection

Observation was taken from 10 plants selected randomly each subplot. Data was recorded on pre and post harvesting stage. The data recorded during pre-harvest; were percentage of germination after 3 weeks and 5 weeks, seeds that failed to germinate were counted per hole (implanting seeds), leaf area index after 90 days. The harvest stage data were: - height of plant, number of primary and secondary branches, and number of pods per plant. The yield of threshed seeds from one meter of each plots were taken and transformed in kg/ha.

## 2.5 Statistical Analysis

The collected data were subjected to standard statistical analysis. The procedure of analysis of variance and mean separation were followed according to the description of [17]. The data was analyzed by MSTAT-C Statistical Package.

## 4. RESULTS AND DISCUSSION

The effect of Jatropha Seed Cake (JSC), sulfur, super phosphate and ammonium sulphate fertilizer on germination % of two groundnut cultivars (Sodri and Gebish) were presented in Table (1a) for season 2011/12 and Table (1b) season 2012/13. There were no significant difference between two cultivars for 3 and 5 weeks of sowing, and among the fertilizers for 5 weeks of sowing, but there were significant differences among fertilizers x cultivars (F×V) for two seasons for 3 and 5 days of sowing. Sulfur fertilizer with V1 (Sodri) and super phosphate with V2 (Gebish) gave the highest values for first and second seasons respectively. Among fertilizers for two seasons in 3 weeks of sowing application of sulfur and (JSC) obtained the highest values. This evaluated that groundnut responded well to application of organic and inorganic fertilizes. This was in line with [9].

The results of leaf area index presented in Table 2, there were significant difference in first season between cultivars, fertilizers and interaction among cultivars and fertilizers.V2, F4 (super phosphate) and the interaction of V2×F4 gave the highest values. While the second season the significant differences were noticed only among the interaction of cultivars and fertilizers and V1×XF4 showed the highest values.

Tables 3a-3b showed the results of effect of JSC and different fertilizers on plant height of two groundnut cultivars for three harvesting periods. The results indicated non significant differences between cultivars except in third harvest of second season and V1 (Sodri) gave the highest value. Among fertilizers treatments all of them revealed highly significant difference except the first harvest of second season. In general, F4 (super phosphate fertilizers)showed the highest records in first harvest of first season and second harvest of second season also with control in second and third harvest of first season, but in second season third harvest F2(JSC) showed the highest value. The interaction of cultivars with fertilizers showed significant difference for all parameters and F4 and F1 obtained the highest values except in the third harvest of second season F2 gave the higher value.

**Table 1a. Effect of Jatropha seed cake and different fertilizers on germination % of two groundnut cultivars season 2011/12**

| | 3 WS | | | 5WS | | |
|---|---|---|---|---|---|---|
| | V1(Sodri) | V2(Gebish) | Means | V1(Sodri) | V2(Gebish) | Means |
| F1(Control) | 70.0b | 69.0b | 69.5ab | 77.5bc | 81.3b | 79.4a |
| F2(Jatropha) | 65.0bc | 64.6bc | 64.8ab | 71.3c | 76.8bc | 74.1a |
| F3(Sulfur) | 82.9a | 58.7c | 70.8a | 89.2a | 69.5c | 79.4a |
| F4(Super ph.) | 66.7bc | 63.8bc | 65.2ab | 75.2bc | 71.7c | 73.4a |
| F5(Amonium sulphate) | 69.6b | 49.6d | 59.6b | 70.6c | 72.8bc | 71.7a |
| Means | 70.8a | 61.1a | | 76.8 a | 74.4a | 75.6 |
| CV% | | | 21.5 | | | 18.9 |
| LSD V | | | 7.8 | | | 7.8 |
| LSD F | | | 10.2 | | | 10.2 |
| LSD V*F | | | 62.7 | | | 54.1 |

**Table 1b. Effect of Jatropha seed cake and different fertilizers on germination % of two groundnut cultivars season 2012/13**

| | 3 WS | | | 5WS | | |
|---|---|---|---|---|---|---|
| | V1(Sodri) | V2(Gebish) | Means | V1(Sodri) | V2(Gebish) | Means |
| F1(Control) | 45.8c | 60.3b | 53.1b | 36.3c | 50.4ab | 43.4a |
| F2(Jatropha) | 54.6b | 60.4b | 59.5a | 46.9b | 50.7ab | 48.8a |
| F3(Sulfur) | 41.9c | 65.4b | 53.7b | 31.7c | 50.5ab | 41.1a |
| F4(Super ph.) | 46.1c | 68.9a | 57.5ab | 37.2c | 56.8a | 47.0a |
| F5(Amonium sulphate) | 57.5b | 57.5b | 57.5ab | 46.4b | 46.4b | 46.4a |
| Means | 49.9a | 62.5a | | 39.7a | 51.0a | |
| CV% | | | 31% | | | 34.6% |
| LSD V | | | 4.86 | | | 11.30 |
| LSD F | | | 9.63 | | | 8.57 |
| LSD V*F | | | 26.7 | | | 15.00 |

**Table 2. Effect of Jatropha seed cake (JSC) and different fertilizers on leaf area index of two groundnut cultivars seasons 2011/12-2012/13**

| | 2011/12 | | | 2012/13 | | |
|---|---|---|---|---|---|---|
| | V1(Sodri) | V2(Gebish) | Means | V1(Sodri) | V2(Gebish) | Means |
| F1(Control) | 29.2efg | 30.6def | 29.9b | 34.7b | 29.3de | 31.9a |
| F2(Jatropha) | 32.3cd | 32.6cd | 32.5ab | 38.1a | 28.7e | 33.4a |
| F3(Sulfur) | 35.2ab | 30.6def | 33.4a | 36.2ab | 31.7cd | 33.9a |
| F4(Super ph.) | 27.89 | 36.4a | 32.1ab | 34.0bc | 28.8e | 31.4a |
| F5(Amonium sulphate) | 28.4fg | 33.8bc | 31.1ab | 34.4b | 30.3de | 32.3a |
| Means | 30.6b | 32.9a | | 35.5a | 29.7a | |
| CV% | 13.2 | | | 14.1 | | |
| LSD V | | | 1.5 | | | 2.5 |
| LSD F | | | 2.3 | | | 3.3 |
| LSD V*F | | | 3.0 | | | 13.7 |

The results showed the effect of JSC and different fertilizers on number of primary branches of two groundnut cultivars seasons 2011/12-2012/13. It was presented in Table 4a-4b for two seasons respectively. The results obtained that there were no significant differences between cultivars for three different harvesting time except in season two second harvest and V1 (Sodri) gave higher record than V2 (Gebish). Among fertilizers in the first season the results showed that F2 (JSC) in first and third harvest and F4 (Super phosphate) gave the highest values with significant differences. Meanwhile, second season showed slight difference in first and second harvest and no significant difference in third harvest for fertilizers treatment. The results of interaction of cultivars with fertilizers showed that F2 for two cultivar in first harvest and V1F4 second harvest and V2 F2 for thirds harvest had the highest values with significant difference. The interaction in season two showed results similar to results of fertilizers with difference in V1which gave the biggest values in first and second harvest, but thirds harvest gave significant difference in V1F4.

Numbers of pods/plants were obtained in Table 5a-5b. The results showed non significant differences between cultivars for two seasons in all harvest time. Fertilizer treatment showed no significant differences in first harvest of first season and second and third harvest of second season. In first season super phosphate gave the significant difference among treatments. In second season F5 gave the highest values. [18] study the relationships between nutrient uptake parameters, biomass and pod dry weight and found that they were positive and significant in both seasons. The interaction treatment revealed highly significant difference for three harvests for two seasons. In general third harvest of two seasons had the highest number of pods/plant and V2F4 of third harvest of second season obtained the heaviest numbers of pods/plant which was 18.3 pods/plant. [19] gave the same result and showed that the genotypes obtained an increasing trend in the number of mature pods as harvesting date delayed.

The yield kg/ha of two groundnut cultivars for three harvests were presented in Table 6a-6b for two seasons. The result showed that yield increased with advanced in harvest time for two seasons, and the yield of third harvest is the best compared with the first and second harvest. This result comes in line with [19] but [20] found that peanut harvest at physiological maturing period gave better result than that at 10 days earlier or later. V2 (Gebish) gave slight big yield compared with V1 (Sodri) with significant difference in first and third harvest of first season and first harvest of second seasons. Among fertilizers treatment in second seasons Tables 6b the results showed that F2 for three harvests gave the highest yield and also for interaction had the same results with variation in cultivars V2F2 gave the highest values. In first season Table 6a, F5 (Amonium sulphate) gave the highest values for three harvest. The interaction of cultivars and fertilizers in this season gave significant difference. The same results among fertilizers treatments were detected [21].

Different concentration of JSC gave the various results on the yield of each vegetable crop. Treatment with the half rate of chemical fertilizer mixed with high rate (10 t/ha) of JSC had proven to provide the best marketable yield of Chinese kale, in which such yield was comparable to that of full rate of chemical fertilizer. In addition, the half rate of chemical fertilizer combined with any rates of JSC (low, medium, and high) gave the most tomato fruit yield. When applied with low rate (2.5 t/ha) of JSC to the sweet potato plot, the outcome showed the highest tuber yield [22]. As mentioned above, fertilizers in form of Jatropha Seed Cake, super phosphate, sulfur and ammonium sulphate on different yield and yield gave positive results with significant difference. [6] Concluded that phosphorus rate increased the number of filled pods and seed yield. Also [23] found that application of FYM and S fertilization increased the yield and yield attributed of groundnut. [24] Confirmed that sufur application increased pod yield of groundnut. On the other hand, [25] found that sulfur application significantly influenced the growth and yield attributed characters, yield and oil content over control regardless of the source and levels of sulfur In contrast [3] found that sulfur supply only increase the sulfur concentration of the plants without enhancing the yield.

**Table 3a. Effect of jatropha seed cake and different fertilizers on plant height of two groundnut cultivars season 2011/12**

| | First harvest | | | Second harvest | | | Third harvest | | |
|---|---|---|---|---|---|---|---|---|---|
| | V1(Sodri) | V2(Gebish) | Means | V1(Sodri) | V2(Gebish) | Means | V1(Sodri) | V2(Gebish) | Means |
| F1(Control) | 18.1cd | 19.8bc | 18.9b | 21.3c | 23.3b | 22.3a | 21.3cde | 26.1a | 23.7a |
| F2(Jatropha) | 20.4b | 17.3d | 18.8b | 19.6def | 20.4cde | 19.9b | 21.9cd | 20.3ef | 21.1b |
| F3(Sulfur) | 18.7bcd | 20.2b | 19.4b | 19.2ef | 22.8b | 21.0ab | 20.1ef | 19.6f | 19.9b |
| F4(Super ph.) | 22.3a | 23.3a | 22.8a | 19.5ef | 24.7a | 22.1a | 22.7bc | 23.6b | 23.2a |
| F5(Amonium sulphate) | 19.5bc | 19.8bc | 19.7b | 18.9f | 21.0cd | 19.9b | 20.7def | 19.8ef | 20.2b |
| Means | 19.8a | 20.1a | | 19.7 a | 22.4a | | 21.3 a | 21.9 a | |
| CV% | 16.4 | | | 11.7 | | | 12.2 | | |
| LSD V | 1.79 | | | 1.35 | | | 1.45 | | |
| LSD F | 2.3 | | | 1.77 | | | 1.89 | | |
| LSD V*F | 3.6 | | | 9.4 | | | 2.7 | | |

**Table 3b. Effect of jatropha seed cake and different fertilizers on plant height of two groundnut cultivars season 2011/12**

| | First harvest | | | Second harvest | | | Third harvest | | |
|---|---|---|---|---|---|---|---|---|---|
| | V1(Sodri) | V2(Gebish) | Means | V1(Sodri) | V2(Gebish) | Means | V1(Sodri) | V2(Gebish) | Means |
| F1(Control) | 19.0d | 22.5bc | 20.8a | 18.1d | 21.6c | 19.9cd | 18.8e | 24.7bc | 21.8ab |
| F2(Jatropha) | 20.5cd | 25.9a | 23.2a | 21.5c | 26.1b | 23.8ab | 21.6d | 27.4a | 24.5a |
| F3(Sulfur) | 23.1b | 22.5bc | 22.8a | 18.6d | 25.9b | 22.3bc | 18.6e | 25.6ab | 22.1ab |
| F4(Super ph.) | 20.4cd | 24.1ab | 22.2a | 23.0c | 29.6a | 26.3a | 23.2cd | 24.1bc | 23.7ab |
| F5(Amonium sulphate) | 21.8bc | 21.7bc | 21.7a | 17.9d | 18.8d | 18.4d | 21.7d | 21.7d | 21.7b |
| Means | 20.9a | 23.3a | | 19.9a | 24.4a | | 20.8b | 24.7a | |
| CV% | 19.2 | | | 21.3 | | | 16.3 | | |
| LSD V | 0.78 | | | 2.52 | | | 2.66 | | |
| LSD F | 2.33 | | | 3.31 | | | 2.03 | | |
| LSD V*F | 3.05 | | | 7.60 | | | 13.7 | | |

**Table 4a. Effect of jatropha seed cake and different fertilizers on number of primary branches of two groundnut cultivars season 2011/12**

| | First harvest | | | Second harvest | | | Third harvest | | |
|---|---|---|---|---|---|---|---|---|---|
| | V1(Sodri) | V2(Gebish) | Means | V1(Sodri) | V2(Gebish) | Means | V1(Sodri) | V2(Gebish) | Means |
| F1(Control) | 4.5bc | 4.1c | 4.5bc | 4.1d | 4.5cd | 4.3c | 5.0bc | 4.8bcd | 4.9ab |
| F2(Jatropha) | 5.6a | 5.3a | 5.6a | 4.7c | 5.2b | 4.9ab | 5.1b | 5.7a | 5.4a |
| F3(Sulfur) | 5.1ab | 3.9c | 5.1ab | 4.7d | 5.2b | 4.9ab | 4.3de | 4.9bc | 4.6b |
| F4(Super ph.) | 4.5bc | 4.2c | 4.5bc | 5.8a | 4.6cd | 5.1a | 4.9bc | 4.5cde | 4.7b |
| F5(Amonium sulphate) | 4.5bc | 4.5 | 4.5bc | 4.4cd | 4.6cd | 4.5bc | 4.8bcd | 4.2c | 4.5b |
| Means | 4.8a | 4.4a | | 4.7a | 4.8a | | 4.8a | 4.8a | |
| CV% | | | 24 | | | 17 | | | 16.5 |
| LSD V | | | 0.61 | | | 0.44 | | | 0.43 |
| LSD F | | | 0.79 | | | 0.58 | | | 0.57 |
| LSD V*F | | | 2.5 | | | 2.6 | | | 1.5 |

**Table 4b. Effect of jatropha seed cake and different fertilizers on number of primary branches of two groundnut cultivars season 2012/13**

| | First harvest | | | Second harvest | | | Third harvest | | |
|---|---|---|---|---|---|---|---|---|---|
| | V1(Sodri) | V2(Gebish) | Means | V1(Sodri) | V2(Gebish) | Means | V1(Sodri) | V2(Gebish) | Means |
| F1(Control) | 6.1a | 4.7cd | 5.4a | 6.2a | 4.3de | 5.3a | 5.0bc | 4.6cd | 4.8a |
| F2(Jatropha) | 5.3b | 3.8e | 4.6b | 4.2e | 3.9e | 4.0b | 5.4abc | 5.2bc | 5.3a |
| F3(Sulfur) | 4.1de | 5.4b | 4.8ab | 5.3bc | 3.8e | 4.6ab | 5.8ab | 5.0bc | 5.4a |
| F4(Super ph.) | 4.5cd | 5.0bc | 4.7ab | 4.9cd | 5.2bc | 5.1a | 6.3a | 4.0d | 5.2a |
| F5(Amonium sulphate) | 6.2a | 4.1de | 5.2ab | 5.7ab | 5.0bcd | 3.3a | 5.8ab | 5.1bc | 5.5a |
| Means | 5.2a | 4.6a | | 5.3a | 4.4b | | 5.7a | 5.7a | |
| CV% | | | 20.2 | | | 24.0 | | | 32.5 |
| LSD V | | | 0.71 | | | 0.83 | | | 0.93 |
| LSD F | | | 0.54 | | | 0.63 | | | 1.21 |
| LSD V*F | | | 2.20 | | | 0.66 | | | 2.31 |

**Table 5a. Effect of jatropha seed cake and different fertilizers on number of pods/plant of two groundnut cultivars season 2011/12**

| | First harvest | | | Second harvest | | | Third harvest | | |
|---|---|---|---|---|---|---|---|---|---|
| | V1(Sodri) | V2(Gebish) | Means | V1(Sodri) | V2(Gebish) | Means | V1(Sodri) | V2(Gebish) | Means |
| F1(Control) | 5.4cd | 8.1a | 6.7a | 8.0b | 05.0d | 06.5b | 13.1bc | 10.9cd | 12.0ab |
| F2(Jatropha) | 7.0abcd | 7.5ab | 7.3a | 5.5cd | 08.1b | 06.8b | 10.6cde | 12.0cd | 11.3b |
| F3(Sulfur) | 5.5cd | 5.3cd | 5.4a | 7.6bc | 08.0b | 07.8ab | 09.6de | 15.8b | 12.7ab |
| F4(Super ph.) | 5.6bcd | 5.0d | 5.3a | 9.0ab | 11.0a | 10.0a | 22.8a | 07.7e | 15.2a |
| F5(Amonium sulphate) | 7.1abc | 5.3cd | 6.2a | 8.7ab | 07.2bcd | 07.9ab | 13.4bc | 09.3de | 11.5ab |
| Means | 6.1a | 6.2a | | 7.8a | 7.8a | | 13.9a | 11.2a | |
| CV% | 53.3 | | | 51.5 | | | 40.7 | | |
| LSD V | 1.80 | | | 2.10 | | | 2.80 | | |
| LSD F | 1.55 | | | 2.20 | | | 3.66 | | |
| LSD V*F | 2.37 | | | 4.10 | | | 7.80 | | |

**Table 5b. Effect of jatropha seed cake and different fertilizers on number of pods/plant of two groundnut cultivars season 2012/13**

| | First harvest | | | Second harvest | | | Third harvest | | |
|---|---|---|---|---|---|---|---|---|---|
| | V1(Sodri) | V2(Gebish) | Means | V1(Sodri) | V2(Gebish) | Means | V1(Sodri) | V2(Gebish) | Means |
| F1(Control) | 05.9d | 09.1b | 07.5bc | 12.5bc | 15.8a | 14.2a | 12.4cd | 17.9a | 15.1a |
| F2(Jatropha) | 07.4abcd | 06.8cd | 07.1c | 11.5c | 14.6ab | 13.1a | 14.4bc | 14.4bc | 14.4a |
| F3(Sulfur) | 08.4bc | 10.9b | 09.7ab | 11.2c | 15.0a | 13.1a | 11.9cd | 14.9b | 13.4a |
| F4(Super ph.) | 10.9a | 07.8bc | 09.3ab | 11.9c | 15.7a | 13.8a | 11.9d | 18.3a | 15.1a |
| F5(Amonium sulphate) | 10.9a | 10.9a | 10.9a | 11.7c | 14.3ab | 12.9a | 15.4b | 10.3d | 12.9a |
| Means | 8.7a | 9.1a | | 11.8a | 15.1a | | 13.2a | 15.2a | |
| CV% | 33.5 | | | 26 | | | 29.3 | | |
| LSD V | 6.60 | | | 2.00 | | | 2.27 | | |
| LSD F | 8.89 | | | 2.63 | | | 2.97 | | |
| LSD V*F | 14.53 | | | 7.10 | | | 23.9 | | |

Table 6a. Effect of jatropha seed cake and different fertilizers on yield (kg/ha) of two groundnut cultivars season 2011/12

| | First harvest | | | Second harvest | | | Third harvest | | |
|---|---|---|---|---|---|---|---|---|---|
| | V1(Sodri) | V2(Gebish) | Means | V1(Sodri) | V2(Gebish) | Means | V1(Sodri) | V2(Gebish) | Means |
| F1(Control) | 250.2 f | 445.7de | 347.9c | 309.3f | 695.5cd | 0502.4d | 0760.5e | 1224.0cde | 0992.3e |
| F2(Jatropha) | 791.2b | 360.3ef | 575.8ab | 942.3cd | 450.8ef | 069.60e | 2675.3ab | 0936.5de | 1805.9c |
| F3(Sulfur) | 967.3a | 455.7cde | 711.5a | 1331.5a | 705.0de | 1018.3b | 2710.5ab | 1621.8c | 2166.2b |
| F4(Super ph.) | 390.7ef | 595.0cd | 492.8bc | 430.0f | 895.2cd | 0662.6c | 1331.0cd | 1569.3c | 1450.2d |
| F5(Amonium sulphate) | 831.0ab | 609.5c | 720.3a | 1119.5bc | 1424.7a | 1272.1a | 2761.1a | 2254.5b | 2507.8a |
| Means | 646.1a | 493.2b | | 826.5a | 834.2a | | 2047.7a | 1521.2b | |
| CV% | | | 50.7% | | | 57.3% | | | 49.5% |
| LSD V | | | 132.0 | | | 2216 | | | 584.7 |
| LSD F | | | 340.9 | | | 572.2 | | | 150.0 |
| LSD V*F | | | 157.9 | | | 260.3 | | | 483.1 |

Table 6b. Effect of Jatropha Seed Cake and different fertilizers on yield (kg/ha) of two groundnut cultivars season 2011/12

| | First harvest | | | Second harvest | | | Third harvest | | |
|---|---|---|---|---|---|---|---|---|---|
| | V1(Sodri) | V2(Gebish) | Means | V1(Sodri) | V2(Gebish) | Means | V1(Sodri) | V2(Gebish) | Means |
| F1(Control) | 124.7e | 126.5e | 193.1d | 327.3d | 0242.8de | 285.1c | 792.3c | 0444.3d | 618.3cd |
| F2(Jatropha) | 062.7f | 677.5a | 370.1a | 089.5f | 1607.7a | 848.6a | 191.2e | 1721.2a | 956.2a |
| F3(Sulfur) | 090.0g | 377.2b | 193.1d | 085.7f | 0713.7b | 399.7b | 142.7e | 1340.5b | 741.6bc |
| F4(Super ph.) | 242.7a | 119.5e | 186.8c | 162.7ef | 0274.0d | 218.3c | 556.5d | 0558.5d | 557.5d |
| F5(Amonium sulphate) | 325.0c | 196.7d | 260.8b | 512.7c | 0432.5c | 472.6b | 719.7c | 0794.8c | 757.3b |
| Means | 152,7a | 299.5b | | 235.6a | 654.19a | | 480.5a | 971.9 a | |
| CV% | | | 38.9% | | | 35.5% | | | 32.6% |
| LSD V | | | 71.8 | | | 666.4 | | | 884.2 |
| LSD F | | | 18.5 | | | 172.1 | | | 228.3 |
| LSD V*F | | | 48.4 | | | 86.4 | | | 129.4 |

## 4. SUMMARY AND CONCLUSION

Groundnut is very important oil seed in the Sudan. The productivity of groundnut depends on proper selection of variety, fertilizer management and other management practices. Currently applications of jatropha seed byproduct as oil diesel remain seed cake waste. This Jatropha Seed Cake can be use as good green manure fertilizers. In this study using Jatropha Seed Cake compared with different inorganic fertilizers; sulfur, superphosphate and ammonium sulphate gave significant difference in growth, yield and yield components of groundnut. On the other hand, delayed harvesting periods gave good results in growth and yield of groundnut. The two cultivars of groundnut obtained similar results without clear variation in their growth. Thus additional studies under more conductive condition are needed to better understand the role of Jatropha Seed Cake plays as fertilizer to increase growth and yield of crops.

## COMPETING INTERESTS

Authors have declared that no competing interests exist.

## REFERENCES

1. Taru VB, Kyagya IZ, Mshelia SI. Profitability of groundnut production in Michika local government area of adamawa state, Nigeria. J. Agric. Sci. 2010;1(1):25-29.

2. ARC (Agricultural Research Corporation) of Sudan. Annual report. Wad-Madani, Sudan; 2003-2010.

3. Gubta UC, Mc Leads JA. Effects of various sources of sulfur on yield and sulfur concentrations of cereals and forages. Can. J. of Soil Sci. 1984;64:169-174.

4. Ashraf M, Athar HR, Harris PJC, Kwon TR. Some prospective strategies for improving crop salt tolerance. Adv. Agro. 2008;97:45-110.

5. DAFF. Groundnut production guide line. Department of Agriculture, Forestry and Fisheries. Pretoria, South Africa; 2010.

6. Shiyam JO, Growth and yield response of groundnut (*Arachis hypogaea* L.) to plant densities and phosphorus on an ultisol in southeastern Nigeria. Libyan Agriculture Research Center Journal. 2012;1(4):211-214.

7. Moraditochaee M. Effects of humic acid, foliar spraying and nitrogen fertilizer management on yield of peanut (*Arachis hypogaea* L.) in Iran. ARPN Journal of Agriculture and Biological Science. 2012;7(4):289-293.

8. Habbasha SF, Taha MH, Jafar NA. Effect of nitrogen fertilizer levels and zinc foliar application on yield, yield attributes and some chemical traits of groundnut. Research Journal of Agriculture and Biological Sciences. 2013;9(1):1-7.

9. Rumbidzai DK, Mabwe C. Response of groundnut (*Arachis hypogeal L*) to inorganic fertilizer use in smallholder farming of Makonde Distric, Zambabwe. Journal of Biology Agriculture and Healthcare. 2014;4(7):78-82.

10. Lombin G, Single L, Yayock JK, A decade of fertilizer research on groundnuts (*Arachis hypogaea* L.) in the Savanna zone of Nigeria. Fertilizer Research. 1985;6(2):157-170.

11. Jamal A, Moon YS, Abdin MZ, Enzyme activity assessment of peanut (*Arachis hypogaea* L.) under slow- release sulfur fertilization. Australian J. of Crop Science, 2010;4(3):169-174.

12. Oniya OO, Bamgboye AI. Production of biodiesel from groundnut (*Arachis hypogaea* L.) oil. Agricultural Engineering International: CIGR Journal. 2014;16(1).

13. Srinophakun P. Prospect of Deoiled *Jatropha curcas* Seedcake as Fertilizer for Vegetables Crops – A Case Study. Journal of Agricultural Science. 2012;4(3):211-226.

14. Keremane BG, Hegde GV, Sheshachar VS. *Jatropha curcas*: Production systems and uses. In: Hegde NG, Daniel JN, Dhar S. (eds.), Proceedings of National Workshop on *Jatropha Curcas* and Other Perennial Oilseed Crops, BAIF Publications, BAIF, Pune, Maharastra; 2003.

15. Openshaw K. A review of *Jatropha curcas*: an oil plant of unfulfilled Promise. Biomass and Bioenergy. 2000;19:1-15.

16. Ganapathy Selvam G, Sivakumar K. Influence of seaweed extract as an organic fertilizer on the growth and yield of (*Arachis hypogaea* L.) and their elemental composition using SEM–Energy Dispersive

Spectroscopic analysis. Asian Pacific Journal of Reproduction. 2014;3(1):18-22.

17. Gomez KA, Gomez AA. Statistical Procedures For Agricultural Research, 2nd edition, A Wily Inter. Sci publication. 6John Wiley and Son, New York. 1984.

18. Junjittakarn J, Pimratch S, Jogoloy S, Htoon W, Singkhan N, Vorasoot N, Tooomsan B, Holbrook CC, Patanothai A. Nutrient uptake of peanut genotypes under different water regimes. International Journal of Plant Production. 2013;7(4):677-692.

19. Kaba JS, Ofori K, Kumaga FK. Inter-relationships of yield and components of yield at different stages of maturity in three groundnuts (*Arashis hypogaea* L) varieties. International J. of Life Sci. Research. 2014;2(1):43-48.

20. Rahmianna AA, Taurfig A, Yusnawan E. Pod yield and kernel quality of peanut grown under two different irrigation and two harvest times. Indonesian Journal of Agriculture. 2009;2(2):103-109.

21. Migawer EA, Soliman MAM. Performance of two peanut cultivars and their response to NPK fertilization in newly reclaimed loamy sand soil. J. Agric. SCi. Mansoura Uni. 2001;26(11):6653-6667.

22. Srinophakun P, Saimaneerat A, Sooksathan I, Visarathanon N, Malaipan S, Charernsom K, Chongrattanameteekul W. Integrated Research on *Jatropha Curcas* Plantation Management. World Renewable Energy Congress. Sweden 8-13 May 2011. Linkoping Sweden. 2011;232-238.

23. Jat RA, Alhawat IPS. Effect of organic manure and sulphur fertilization in pigeon pea (*Cajanus cajan*) and groundnut (*Arachis hypogaea* L.) intercropping system. International Crops Research Institute for the Semi-Arid Tropics (ICRISAT) Hyderabad, India. 2010;55(4):276-281.

24. Dash AK, Nayak BP, Panigrany N, Mohapatra S, Samant PK. Performance of groundnut (*Arachis hypogaea*, L.) under different levels of sulphur and irrigation. Indian J. of Agronomy. 2014;58(4):578-582.

25. Tajeswara KR, Upendra Rao A, Sehha D. Effect of source and levels of sulfur on groundnut. Journal of Academic and Industrial Research. 2013;2(5):268-274.

# 13

# The Quality of Jute Mallow Seeds Exposed to Different Hot Water-Steeping and Cooling Protocols

**K. D. Tolorunse[1*], H. Ibrahim[1], N. C. Aliyu[1] and J. A. Oladiran[1]**

[1]Department of Crop Production, School of Agriculture and Agricultural Technology, Federal University of Technology, P.M.B. 65, Minna, Niger State, Nigeria.

*Authors' contributions*

*This work was carried out in collaboration between all authors. Author KDT design the study, wrote the protocol and wrote the first draft of the manuscript. Author JAO reviewed the experimental design and all drafts of the manuscript. Authors HI and NCA managed the analyses of the study. Authors HI and NCA identified the seeds. Authors KDT and JAO performed the statistical analysis. All authors read and approved the final manuscript.*

Editor(s):
(1) Lanzhuang Chen, Laboratory of Plant Biotechnology, Faculty of Environment and Horticulture, Minami Kyushu University, Miyazaki, Japan.
(2) Moreira Martine Ramon Felipe, Departamento de Enxeñaría Química, Universidade de Santiago de Compostela, Spain.
Reviewers:
(1) Anonymous, Pakistan.
(2) Abdelaziz Zahidi, Department of Biology, University Ibn Zohr Agadir, Morocco.
(3) Anonymous, USA.
(4) Anonymous, Malaysia.

## ABSTRACT

Steeping of dormant jute mallow (*Corchorus olitorius* L.) seed in hot water at high temperature for enhanced germination, seems to be the most favoured of all other methods. Literature however, appears to be silent on the cooling protocol to adopt to ensure high quality after a seed lot may have been steeped in hot water. Seeds of two cultivars of this crop were subjected to nine hot water/cooling treatments and then dried back. They were then thinly spread and stored in open glass dishes at 83% relative humidity and 33°C. The moisture content of seeds steeped in water at 80 and 97°C increased from about 5-6% prior to storage to about 10-11% after 6-18 weeks after storage (WAS). Steeping of seeds at 80 and 97°C for 5 seconds significantly enhanced germination to about 88% and 77% in 'Amugbadu' and 'Oniyaya' respectively compared to about 8% in the control (unsteeped) seed. Cultivar 'Amugbadu' seeds steeped in cold water (*ca* 27°C) immediately after steeping in water at 97°C recorded higher germination percentage of 90% - 46% within 0-12 weeks of storage compared to the range of 88 to 29% recorded within the same period in seeds that

*Corresponding author: E-mail: kehindetolorunse@yahoo.com;*

were simply left to gradual cooling in ambient condition. 'Oniyaya' seeds exhibited no differential response to cooling protocol. Unsteeped (control) seed of both cultivars recorded higher germination of 82/85% at 20 WAS than seeds steeped at 97°C (irrespective of cooling protocol) and 80°C prior to storage. Furthermore, whereas seeds of the latter group germinated slower and less uniformly as storage progressed, a gradual increase in the values of these vigour indices were recorded in the former group of seeds. Across steeping treatments the germination percentage of 'Oniyaya' seed declined less rapidly than that of 'Amugbadu' seeds during storage from *ca* 61% to 32% in the former compared with about 62% to 3% in the latter.

Keywords: *Corchorus olitorius; seed dormancy; steeping; storage; germination rate and synchronization.*

## 1. INTRODUCTION

Dormancy poses a serious problem to successful seed germination and seedling establishment in jute mallow and different methods have been employed to alleviate the problem. [1] reported the effectiveness of hot water-steeping in this respect which has been confirmed by other reports [2,3,4]. [5] recorded best germination when seeds of *Corchorus olitorius* were first pre-chilled at 6°C for 3½ days before they were incubated at 35°C. Exposure of the seed of this crop to dry heat at 90°C for 5 or ten minutes has also been reported to break dormancy by [6]. According to [7], the temperature required to break seed dormancy in *C. cunninghamii* in Australia ranged between 80-100°C. It was therefore concluded that soil heating from bush fire would promote the germination of seeds of this species. [4,8] reported the effectiveness of sulphuric acid and seed coat scarification respectively. Earthworm cast leachate has also been found effective in dormancy breaking in *C. olitorius* seeds [2]. Of all the methods stated above, though not an exhaustive list, the use of hot water appears to be the most favoured as water is cheaply available and can be more safely handled by all users.

The usual practice when hot water is used to break seed dormancy in this crop is to immediately steep treated seeds in cold water which is assumed to bring down seed temperature and so reduce damages that may occur to the embryo by the treatment at high temperature. [9] recorded better germination and seedling growth when maize seeds that had been stored in deep freezer and refrigerator were first left to equilibrate in ambient condition for some days before they were tested at a temperature higher than that in which they were stored. The usual practice when hot water is used to break seed dormancy in this crop is to immediately steep treated seeds in cold water to bring down seed temperature. It is of the opinion that this would reduce damages that may occur to the embryo by the treatment at high temperature. It is possible that exposure of jute mallow seeds to extreme temperatures in quick successions may damage the embryo. This study aimed at determining the effect hot water-steeping/cooling protocol may have on seed quality in two different cultivars of *C. olitorius* in a bid to enhance seed germination.

## 2. MATERIALS AND METHODS

Samples of *Corchorus olitorius* seeds of cultivars 'Oniyaya' and 'Amugbadu' harvested in November, 2012 were subjected to nine treatments: St1) seeds steeped in hot water at 97°C for 5 seconds and immediately spread out on absorbent paper to cool down and dry in ambient condition (temperature of about 30°C and relative humidity of about 30%); St2) seeds steeped as in St1 followed immediately by steeping in cold water as recommended by [1] as being optimum for obtaining high germination of dormant *Corchorus olitorius* seeds as well as an improvement in seedling emergence; St3) seeds steeped as in St1 followed immediately by steeping in hot water at 80°C; St4) seeds steeped as in St1 followed immediately by steeping in hot water at 60°C; St5) seeds steeped as in St1 followed immediately by steeping in warm water at 40°C; St6) seeds steeped in hot water at 80°C followed immediately by steeping in cold water; St7) seeds steeped in hot water at 60°C followed immediately by steeping in cold water; St8) seeds steeped in warm water at 40°C followed immediately by steeping in cold water; St9) control (non-treated).

Seeds of the different treatments were left to dry on absorbent paper in ambient condition for seven days. The moisture content at which seed equilibrated with the ambient condition was then determined using the high constant temperature oven method [10]. This involved the drying of two replicates of seeds of about 1g each in the oven at 130°C for 1hr. After this period seeds and container were cooled over silica gel in a desiccators for 1hr. The weight of dry seeds was subsequently taken and the moisture content was determined on wet weight basis. Samples of each of the treatments were thinly spread in glass Petri dishes and placed in an incubator at 33±0.5°C and at 83±2.5% for a period of 20 weeks. Seed samples were drawn for germination test at two-weekly interval. In addition to the nine treatments listed above, samples of the untreated (St9-control) seeds were steeped as in St2 above following storage for 12-20 weeks to obtain St10. This was to ascertain the viability of the untreated seeds at these points. Samples of St7 and 8 were similarly tested at 18 and 20 WAS to obtain St11 and 12 respectively. On each of the testing days, four replicates of 50 seeds each were counted and spread on distilled water-moistened filter paper and incubated at 30°C for 16 days. Germination counts were taken every-other-day and the final cumulative figures were expressed as a percentage of the total seed incubated for each treatment. To further index seed quality, mean germination time was determined using the expression recommended by [11]:

$$t = \frac{\sum ni \times ti}{\sum n} \text{ (days)}$$

Where:

t-   Mean germination time,
ti-  Given time interval,
ni-  Number of germinated seeds during a given time interval
n-   Total number of germinated seeds.

Germination synchronization was determined using the formula adopted by [12]: $Z = \sum C_{ni, 2} / N$, with $C_{ni, 2} = n_i(n_i-1)/2$ and $N = \sum n_i (\sum n_i-1)/2$

Where $Cn_{i, 2}$: combination of the seeds germinated in the time $i$ and $n_i$: number of seeds germinated in the time $i$. $Z = 1$ when the germination of all seeds occur at the same time and $Z = 0$ when at least two seeds could germinate, one at each time. The higher the value obtained, the better synchronised the germination.

Seed moisture content for St1-9 were again tested at 6 and 18 weeks of storage to determine if there had been changes to the moisture levels

compared to what was recorded at the beginning of storage.

Seed germination data (in percentages) were transformed to arcsin values before they were statistically analysed. All data were analysed using the SAS statistical package [13]. The least significant difference method (LSD) and Duncan Multiple Range Test were used for mean separation where significant differences were obtained.

## 3. RESULTS

Prior to storage (0WAS), moisture contents of St1-9 ranged between 4.8-6.1 and 4.6-6.4% for "Amugbadu" and "Oniyaya" respectively (Table 1) and by six weeks after storage (6WAS), the values had risen considerably to about 10.5% in all seeds steeped in hot water at 97 and 80°C (i.e. St1-6) with little or no changes at 18WAS. Relatively slight moisture increases were recorded in untreated seeds and in those steeped at 40 and 60°C. Steeping of seeds at 97°C (irrespective of cooling protocol) and at 80°C (St1-6) significantly enhanced germination from about 8% in the control to about 68-88% across variety prior to storage i.e at 0WAS (Table 2). Though seeds steeped in hot water at 60°C generally germinated higher than those steeped at 40°C and the control (especially for 'Amugbadu' from 0-14WAS), the maximum values of 27% and 13% for Amugbadu at 2WAS and 'Oniyaya' at 14WAS respectively were still low. Fig. 1 shows that seed germination was significantly higher in 'Amugbadu' (62%) than in 'Oniyaya" (51%) across steeping treatments at 0WAS. By 2WAS the germination of 'Oniyaya' seeds had improved to about 60% and was statistically similar to that of 'Amugbadu' up to 6WAS. The interaction between steeping treatment and cultivar on seed germination was only significant from 8WAS (Table 2). It is obvious that seeds of 'Oniyaya' steeped at 80 and 97°C germinated significantly higher than those of 'Amugbadu' from 8 to 20WAS. For example, whereas 'Oniyaya' seeds of St1-6 recorded germination values of 71% - 89% at 8WAS, those of 'Amugbadu' had 49% - 72%. At 20WAS ranges of 36% - 56% and 1% - 7% were recorded for 'Oniyaya' and 'Amugbadu' respectively. Germination percentages remained at fairly high levels (63%-79%) in 'Oniyaya' for about 12/14WAS in St1-6 (Table 2) whereas this level of germination was not obtained beyond about 6/8WAS in 'Amugbadu'; generally, no significant differences were recorded between cultivars in St7-9. Table 2 shows further that the cooling protocol adopted following steeping of seeds in hot water at 97°C affected the longevity of seeds of the two cultivars differently. 'Amugbadu' seeds that were steeped in cold water immediately after hot water exposure (St2) germinated significantly higher (65% - 46%) at 8 to 12WAS than seeds that were left to cool gradually under ambient condition (St1) with 57% - 29% germination. Seeds that were steeped at 80, 60 and 40°C following steeping at 97°C before cold water-steeping (St3-5) were only significantly lower in germination than St1 seeds at 8WAS. Seeds steeped at 80°C only, followed immediately by cold water-steeping (St6) survived better (72%-7%) than St1 seeds (with 57% - 1% germination) during 8-20WAS. In cultivar 'Oniyaya' there was generally no differential response to cooling protocol in seeds steeped at 97°C. However, seeds of this cultivar that were steeped at 80°C only (St6) germinated significantly higher (56% - 63%) than St1 (40%-35%) at 16-20WAS. When seeds of both varieties from the control treatment (St9) were steeped in hot water at 97°C for 5 seconds (St10) following 12-20 weeks of storage germination was greatly enhanced to about 78-89% (Table 3). Also, 'Amugbadu' seeds previously subjected to 40 and 60°C steeping prior to storage (St8 and 7 respectively in Table 3) and which germinated poorly (1%-16%) throughout the storage period, had improved germination of 72% - 84% when subjected to hot water steeping at 97°C for 5 seconds at 18 and 20WAS (St12 and 11 respectively in Table 3). However, "Oniyaya" seeds that were previously subjected to 60°C before they were again steeped at 97°C (i.e. St12), germinated significantly lower (59 and 39% at 18 and 20WAS respectively) than seed of St 10 and 11.

Germination took significantly longer (*ca* 9.5 days) at the onset of storage in the unsteeped seeds (St9) than in those steeped in hot water (*ca* 1.6–2.7days) while all hot water-treated seeds germinated at statistically similar rates irrespective of cooling protocol (Fig. 1). As from 2WAS, germination rate of St9 (control) seeds generally improved till the end of storage (from about 9.5 days to about 1.0 day). This trend was also recorded for St5 seeds from 14-20WAS. The effect of cultivar on the germination percentage was significant all through the storage period except at 4 and 6 WAS. Seed of cultivar Oniyaya germinated significantly higher than those of Amugbadu except at 0 and 2 WAS (Fig. 2). Fig. 3 shows that for all the seeds

steeped at 97°C (irrespective of cooling protocol), as well as those steeped at 80°C, germination was consistently fast up to 4WAS taking only about 2-2.7 days, followed by a steady decline as from 6WAS. Fig. 4 reveals that seeds of the two varieties were statistically similar in germination rate within the first 10 weeks of storage. As from 12WAS however, the superiority of "Oniyaya" over "Amugbadu" became significant; while the mean germination time for former ranged between about 2.5 – 3.0 days that of the former was about 3.8 – 4.5days. Fig. 5 shows that prior to storage (i.e. at 0WAS), seeds generally germinated more uniformly when steeped in hot water at 97°C irrespective of the cooling protocol adopted (St1-5) and when steeped at 80°C (St6) with values ranging from about 0.7 to 0.9 than when they were steeped at 40 and 60°C (St7 and 8 with value of about 0.2) or when unsteeped (St9 with about 0.1 synchronization value). As storage progressed however, whereas germination became more non-uniform in St1-6, there were gradual improvements in St7-9. Germination was better synchronized in "Oniyaya" with value of about 0.3 than in "Amugbadu" with values of about 0.5 – 0.6 at 12, 14 and 16WAS (Fig. 6).

**Table 1. Percentage moisture content (on wet weight basis) of seeds of the various treatments at the onset and after 6 and 18 weeks of storage at 83% relative humidity and 33°C**

| Cultivar | Steeping treatment | Storage period (weeks) | | |
|---|---|---|---|---|
| | | 0 | 6 | 18 |
| 'Amugbadu' | St1 | 5.5 | 9.9 | 9.4 |
| | St2 | 6.0 | 10.4 | 10.4 |
| | St3 | 5.8 | 10.6 | 9.5 |
| | St4 | 5.6 | 9.5 | 10.0 |
| | St5 | 6.1 | 10.4 | 10.0 |
| | St6 | 5.9 | 10.0 | 10.0 |
| | St7 | 5.1 | 7.0 | 6.1 |
| | St8 | 5.3 | 5.6 | 5.1 |
| | St9 | 4.8 | 5.4 | 5.5 |
| 'Oniyaya' | St1 | 4.6 | 10.5 | 11.3 |
| | St2 | 5.0 | 10.6 | 11.0 |
| | St3 | 5.9 | 10.6 | 10.7 |
| | St4 | 6.3 | 10.1 | 10.3 |
| | St5 | 6.1 | 9.7 | 10.4 |
| | St6 | 6.4 | 10.2 | 9.8 |
| | St7 | 4.7 | 5.4 | 5.0 |
| | St8 | 5.3 | 5.5 | 5.7 |
| | St9 | 5.8 | 4.9 | 5.6 |

**Table 2. Interaction effect of cultivar and steeping treatment on the germination percentage of *Corchorus olitorius* seeds at 8-20WAS**

| Cultivar | Steeping treatment | Storage period (weeks) | | | | | | | | | | |
|---|---|---|---|---|---|---|---|---|---|---|---|---|
| | | 0 | 2 | 4 | 6 | 8 | 10 | 12 | 14 | 16 | 18 | 20 |
| 'Amugbadu' | St1 | 86 | 82 | 82 | 85 | 57de | 40c | 29c | 13de | 5def | 3d | 1ef |
| | St2 | 86 | 90 | 87 | 89 | 65bc | 57b | 46b | 12de | 11cde | 7d | 3def |
| | St3 | 87 | 83 | 84 | 90 | 52e | 54bc | 37bc | 10de | 8cdef | 5d | 1ef |
| | St4 | 88 | 89 | 84 | 86 | 52e | 49bc | 38bc | 12de | 15cd | 5d | 4def |
| | St5 | 88 | 91 | 88 | 84 | 49e | 51bc | 39bc | 13de | 7cdef | 2d | 1ef |
| | St6 | 84 | 87 | 89 | 84 | 72ab | 58b | 46b | 32c | 18c | 16c | 7cd |
| | St7 | 19 | 27 | 15 | 12 | 15f | 16d | 9d | 24cd | 5def | 4d | 6cd |
| | St8 | 8 | 8.5 | 3 | 7 | 4g | 1f | 4de | 4ef | 1f | 1e | 3def |
| | St9 | 8.5 | 6 | 7 | 4 | 5g | 6ef | 4de | 1fg | 1f | 4d | 4def |
| 'Oniyaya' | St1 | 68 | 86 | 84 | 87 | 84a | 79a | 75a | 66ab | 38b | 40b | 35b |
| | St2 | 77 | 88 | 86 | 88 | 89a | 84a | 75a | 63b | 47ab | 45ab | 61a |
| | St3 | 73 | 87 | 88 | 90 | 71abc | 85a | 71a | 66ab | 33b | 52a | 34b |
| | St4 | 76 | 88 | 80 | 87 | 86a | 88a | 72a | 77a | 50a | 57a | 52a |
| | St5 | 70 | 90 | 87 | 93 | 84a | 85a | 74a | 79a | 49a | 41b | 33b |
| | St6 | 77 | 82 | 86 | 84 | 81a | 83a | 74a | 77a | 53a | 50a | 56a |
| | St7 | 8 | 12 | 7 | 10 | 6fg | 8de | 6de | 13de | 1f | 2e | 13c |
| | St8 | 3 | 6 | 5 | 3 | 2g | 1f | 3e | 0g | 1f | 2d | 1f |
| | St9 | 8 | 4 | 3 | 7 | 2g | 5ef | 3e | 6ef | 3ef | 4d | 3def |
| | Significance | Ns | ns | ns | ns | * | * | * | * | * | * | * |

Means followed by the same letter in a column are not significantly different at P≥ 0.05 using the Duncan's Multiple Range test.

St1 = steeped at 97°C for 5 seconds and air-cooled at room temperature; St2 = steeped at 97°C for 5 seconds and immediately steeped in cold water(27 oC); St3 = steeped at 97°C for 5 seconds followed by 5 seconds steeping at 80°C, followed by steeping in cold water(27 oC); St4 = steeped at 97°C for 5 seconds followed by 5 second steeping at 60°C followed by steeping in cold water(27 oC); St5 = steeped at 97°C for 5 seconds followed by 5 seconds steeping at 40°C followed by steeping in cold water(27 oC); St6 = steeped at 80°C for 5 seconds followed by steeping in cold water (27 oC); St7 = steeped a 60°C and immediately steeped in cold water(27 oC); St8 = steeped at 40°C for 5 seconds and immediately steeped in cold water(27 oC); St9 = control (no hot water steeping)

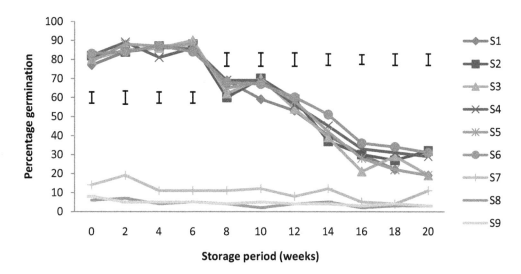

**Fig. 1. Effect of steeping treatment on seed germination percentage following storage**
*I: LSD bar at P=0.05*

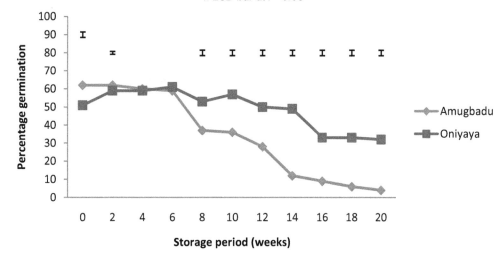

**Fig. 2. Effect of variety on seed germination percentage following storage**
*I: LSD bar at P=0.05*

**Table 3. Germination percentages of previously untreated seeds that were subsequently steeped at 97°C for 5 sec. (St10) and those previously steeped at 40 and 60°C prior to storage that were subsequently steeped at 97°C for 5 sec. (St11 and 12 respectively) following storage for the periods indicated**

| Cultivar | Steeping treatment | Storage period (weeks) | | | | |
|---|---|---|---|---|---|---|
| | | 12 | 14 | 16 | 18 | 20 |
| 'Amugbadu' | St10 | 84a | 86a | 78a | 83ab | 85a |
| | St11 | - | - | - | 84ab | 72b |
| | St12 | - | - | - | 72b | 77ab |
| 'Oniyaya' | St10 | 89a | 82a | 80a | 82ab | 82a |
| | St11 | - | - | - | 90a | 84a |
| | St12 | - | - | - | 59c | 39c |

*Means followed by the same letter in a column are not significantly different at P≤ 0.05 using the Duncan's Multiple Range test*

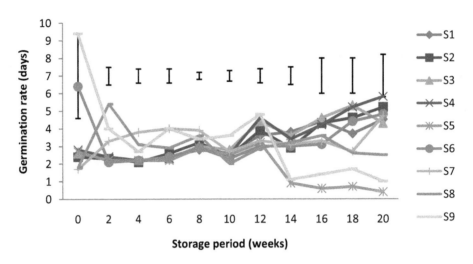

**Fig. 3. Effect of steeping treatment on seed germination rate (days) following storage**
*I: LSD bar at P=0.05*

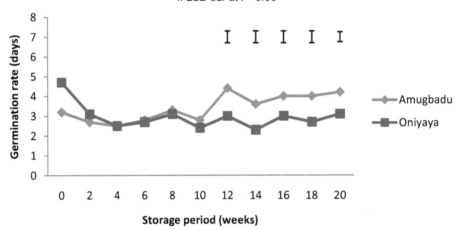

**Fig. 4. Effect of variety on seed germination rate (days) following storage**
*I: LSD bar at P=0.05*

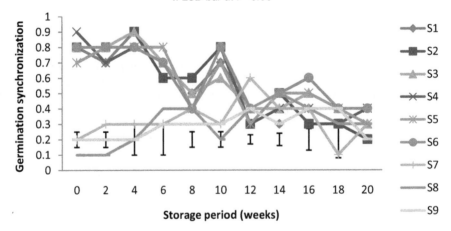

**Fig. 5. Effect of steeping treatment on seed germination synchronization following storage**
*I: LSD bar at P=0.05*

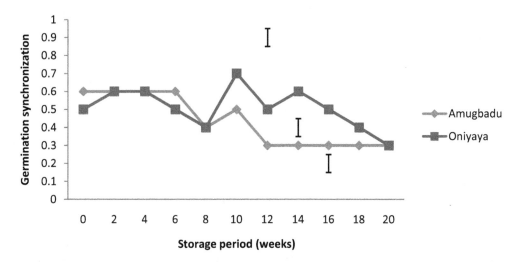

## Fig. 6. Effect of variety on seed germination synchronization following storage
*I: LSD bar at P=0.05*

## 4. DISCUSSION

The considerable increase in the moisture content of seeds of St1-6 during storage at high relative humidity as recorded in this study is an indication of improved permeability due to the softening of the coats by hot water. It is this change in permeability that must have been responsible for the loss of dormancy (and therefore, enhanced germination), a confirmation that dormancy in this species is physical caused by water-impermeable seed coat [14]. [1] had hypothesized that hot water may be alleviating dormancy in *Corchorus* by weakening the seed coat. [2] postulated that some chemicals secreted by earthworm are also capable of weakening woody seed coat of jute seed. Steeping in water at 40 and 60°C did not result in substantial seed coat permeability and therefore the seed remained relatively dry (with moisture content of an average of about 5-6%) and retained high level of dormancy. [6] have also reported the ineffectiveness of dry heat of 40-60°C compared to the effectiveness of 90°C in breaking *Corchorus olitorius* seed dormancy. The effectiveness of temperatures of 80 and 97°C in overcoming seed dormancy as recorded here, agrees with that reported by [7] for *C. cunninghamii*. Based on [7] report, [15] asserted that temperature of 60-70°C which may be produced by low intensity forest fire may not be sufficient to stimulate *C. cunninghamii* seed germination. However, the ability of some seeds exposed to 60°C to significantly germinate (though low) better than the untreated ones, whereas most seeds germinated well following

exposure to 80-97°C in the current study, is indicative of variability in the depth of dormancy of the individual *C. olitorius* seeds. This view agrees with that expressed by [16] for *Phloxpilosa* seeds. The existence of differential rates of seed dormancy release in *C. cunninghamii* was alluded to in the report by [14]. Variability of dormancy depth is seen as a survival strategy [17] which ensures the perpetuation of species. The poorer longevity of 'Amugbadu' seeds that were subjected to gradual cooling in ambient condition following hot water-steeping compared with those that were steeped in cold water immediately, indicates that hot water steeping at 97°C must have resulted in embryo damage and that the extent of such damage could be reduced by immediately steeping treated seeds in cold water.

The poor germination of 'Oniyaya' seeds steeped in hot water at 60°C compared with the values obtained for the untreated seeds that were subsequently subjected to steeping at 97°C at 18 and 20WAS, is an indication that the earlier exposure to 60°C (despite immediate steeping in cold water) damaged the seed embryo even when the treatment did not result in considerable dormancy alleviation. The greatest damaged to seed longevity was occasioned by steeping of seed at 80-97°C. The poorer storability of hot water-steeped seeds as recorded in this study is in agreement with recent report by [18] which revealed that potential seed longevity was adversely affected even by 2-second hot water-steeping. [19] also reported lower germination value for seeds stored following hot water steeping in comparison to unsteeped seeds, though the difference between the two seed lots was insignificant. That treated 'Oniyaya' and 'Amugbadu' seeds still gave acceptable germination percentages at 12/14 (up to 79%) and 6WAS (up to 90%) respectively despite the adverse effect of hot water steeping on potential seed longevity in the two cultivars used in this study is note worthy. This result runs contrary to the report by [20] that hot water steeped seeds cannot be stored. Better longevity of such seeds will even be attained if storage is in more conducive environment of lower humidity and temperature. The significant variation in seed longevity of the two jute varieties in this study agrees with the view expressed [21] that this trait may be species or variety specific. Differences in seed longevity among varieties have also been reported in other crops [22,23,24]. The poorer germination of 49 – 65% recorded as from 8WAS for 'Amugbadu' seeds that were previously steeped at 97°C compared with 71–89% for 'Oniyaya' is an indication of faster deterioration in the former. Differences in genotype may also explain the significant steeping treatment x cultivars interaction effect on germination percentage recorded from 8WAS. It has been established from this study that Amugbadu seeds are poor storers. Their poor vigour which became evident from 8 WAS must have been responsible for their inability to withstand hot water steeping stress compared to Oniyaya seeds. Ability to tolerate stressful condition is known to vary with cultivars [25].

The enhanced total germination, germination rate and synchronization before storage due to hot water-steeping can be attributed to improved seed coat permeability. [26] stated that seed treatments may overcome dormancy and enhance germination by altering the physical integrity of seed coverings. This alteration could also constitute damage to the seed coat. Damage to seed coat has been reported to result in seed deterioration in soybean and also reducing storage ability and germination rate [27]. Subsequent decline in performances in respect of total germination, germination rate and synchronization as from about 8WAS is indicative of decline in seed vigour. However, whereas a decline in seed vigour is normally known to occur before loss of germination [28], the two occurred around the same time (ca 8WAS) in this study for seeds exposed to 80-97°C.

According to [29] both dormancy and germination are influenced by the combined effects of the potential growth of the embryo and the resistance of the surrounding tissues. The significantly better germination of seeds of 'Amugbadu' (84–88%) than 'Oniyaya' (68–77%) at 0WAS following steeping at 80 and 90°C, with substantial improvement to about 90% in the latter as storage progressed as recorded in this study may therefore, suggest that the coats of 'Oniyaya' seed still placed significant restraint on the embryo even after steeping and that this was relaxed by 2WAS. The subsequent decline in the percentage germination and vigour, indexed using germination rate and synchronization of the seeds of both cultivars could be adduced to decline in embryo potential growth. Aging has also been reported to result in delayed and decreased germination in other crops [30]. Contrary to the trend above, the improvement in seed germination rates and synchronization recorded for untreated seeds and those exposed to only 40 and 60°C steeping as seeds aged could be assumed to be due to the weakening of the seed coverings as storage progressed. According to [31] germination speed is influenced by a co-action of the embryo's growth potential and a reduction in the physical strength of surrounding seed coverings.

## 5. CONCLUSION

It is concluded that seeds of the two cultivars used in this study possessed dormancy which was broken by steeping in water at 80–97°C for five seconds. The treatment also resulted in faster and more synchronized germination compared with the control. The study also revealed that it may be necessary to immediately steep hot water treated seeds in cold water at 27°C for better longevity. Seeds of cv 'Oniyaya' retained higher germination for longer (ca 12/24 WAS) than those of cv 'Amugbadu' (about 6WAS). Seed germination percentage, germination rate and synchronization declined from about 8WAS to the end of the storage period.

## COMPETING INTERESTS

Authors have declared that no competing interests exist.

# REFERENCES

1. Oladiran JA. Effect of stage of harvesting and seed treatment on germination, seedling emergence and growth in *Corchorus olitorius* 'Oniyaya'. Scientia Horticulturae. 1986;28:227-233.

2. Ayanlaja SA, Owa SO, Adigun MO, Senjobi BA, Olaleye AO. Leachate from earthworm castings breaks seed dormancy and preferentially promotes radical growth in jute. Hort Science. 2001;36(1):143-144.

3. Velempini P, Riddoch I, Batisani N. Seed treatment for enhancing germination of wild okra (*Corchorus olitorius*). Experimental Agriculture. 2003;39:441-447.

4. Emongor VE, Mathowa T, Kabelo S. The effect of hot water, sulphuric acid, nitric acid, gibberellic acid and ethephon on the germination of Corchorus (*Corchoru stridens*) seed. Journal of Agronomy. 2004;3:196-200.

5. Nkomo M, Kambizi L. Effect of pre-chilling and temperature on seed germination of *Corchorus olitorius* L. (Tilliaceae) (Jews mallow), a wild leafy vegetable. African Journal of Biotechnology. 2009;8(6):1078-1081.

6. Denton OA, Oyekale KO, Adeyeye JA, Nwangburuka CC, Wahab OD. Effect of dry heat treatment on the germination and seedling emergence of *Corchorus olitorius* seed. Agricultural Science Research Journals. 2013;3(1):18-22.

7. Halford D. *Corchorus cunninghamii*, A conservation assessment, Report to the Australian Nature Conservation Agency, Project No. 1993;317.

8. Chauhan B, Johnson DE. Seed germination and seedling emergence of Nalta jute (*Corchorus olitorius*) and Redweed (*Melochia concatenate*): Important broadleaf weeds of the tropics. Weed Science. 2008;56(6):814-819.

9. Ajayi SA, Fakorede MAB. Effect of storage environments and duration of equilibration on maize seed testing and seedling evaluation. Maydica. 2001;46:267-275.

10. ISTA. International rules for seed testing. Edition 2005.International Seed Testing Association. Bassersdorf, Switzerland; 2005.

11. Labouriau LG. A germinacao das sementes. Organizacao dos Estados Americanos. Programa Regional de Desenvolvimento Cientifico e Tecnologico Serie de Biologia. Monografia. 1983;24.

12. Ranal MA, Santana DG. How and why to measure the germination process? Revista Brasileira de Botanica. 2006;29(1):1-11.

13. Statistical Analysis software (SAS).System for windows. SAS User's Guide; Statistics, Version 9.1.SAS Institute Inc. Cary. NC, USA. 1999;1028.

14. Baskin JM, Baskin CC, Li X. Taxonomy, anatomy and evolution of physical dormancy in seeds. Plant Species Biology. 2000;15:139-152.

15. NSW Department of Environment and Conservation. Draft Recovery Plan for *Corchorus cunninghamii*, NSW Department of Environment and Conservation, Hurstville. 2004;17.

16. Madeiras AM, Boyle TH, Autio WR. Germination of *Phlox pilosa* L. seeds is improved by gibberellic acid and light but not scarification, potassium nitrate, or surface disinfestations. Hort Science. 2007;42(5):1263-1267.

17. Stewart JR, McGary I. Brief exposure to boiling water combined with cold-moist stratification enhances seed germination of New Jersey tea. Hort Technology. 2010;20(3):623-625.

18. Tolorunse KD, Ibrahim H, Oladiran JA. Hot water-steeping enhanced germination of jute mallow (*Corchorus olitorius*) seeds. Advanced Crop Science. 2013;8:542-548.

19. Ibrahim H, Oladiran JA, Mohammed H. Effects of seed dormancy level and storage container on seed longevity and seedling vigour of jute mallow (*Corchorus olitorius*). African Journal of Agricultural Research. 2013;8(16):1370-1374.

20. Schippers RR. African indigenous vegetables. An overview of the cultivated species. Chatham, UK. Natural Resources Institute/ACP-EU Technical Centre for Agricultural and Rural Cooperation; 2000.

21. McDonald MD. Seed deterioration: physiology, repair and assessment. Seed Science and Technology. 1999;27:177-237.

22. Kamaswara RM, Jackson MT. Seed production environment and storage longevity of Japonica rice (*Oryza sativa* L.). Seed Science Research. 1996;6:17-21.

23. Mrda J, Crnobarac J, Dusanic N, Radic V, Miladinovic D, Jocic S, Miklic V. Effect of storage period and chemical treatment on sunflower seed germination. HELIA. 2010;33(53):199-206.

24. Adebisi MA, Abdul-Rafiu AM, Abdul RS, Daniel IO, Tairu FM. Seed longevity and vigour of watermelon (*Citrillus lanatus* Thumb Mansf.) seeds stored under ambient humid tropical conditions. Nigerian Journal of Horticultural Science. 2012;17:169-176.

25. Aboutalebian MA, Mohagheghi A, Niaz SA, Rouhi HR. Influence of hydropriming on seed germination behaviour of canola cultivars as affected by saline and drought stresses. Annals of Biological Research. 2012;3(11):5216-5222.

26. Katzman LS, Taylor AG, Langhans RW. Seed enhancements to improve spinach germination. Hort Science. 200136(5):979-981.

27. Okabe A. Inheritance of seed coat cracking and effective selection method for resistance in soybean. Japan Agricultural Research Quarterly. 1996;30:15-20.

28. Demir I, Mavi K. Controlled deterioration and accelerated aging tests to estimate the relative storage potential of cucurbit seed lots. Hort Science. 2008;43(5):1544-1548.

29. Koorneef M, Bentsink L, Hilhorst H. Seed dormancy and germination. Current Opinion in Plant Biology. 2002;5:33-36.

30. Hacisalihoglu G, Taylor AG, Paine DH, Hilderbrand MB, Khan AA. Embryo elongation and germination rates as sensitive indicators of lettuce seed quality: priming and aging studies. Hort Science. 1999;34(7):1240-1243.

31. Welbaum GE, Bradford KJ, Yim KO, Booth DT, Oluoch MO. Biophysical, physiological and biochemical processes regulating seed germination. Seed Science Research. 1998;8:161-172.

# 14

# Effects of Polybag Size and Seedling Age at Transplanting on Field Establishment of Cashew (*Anacardium occidentale*) in Northern Ghana

Patricia Adu-Yeboah[1*], F. M. Amoah[1], A. O. Dwapanyin[1],
K. Opoku-Ameyaw[1], M. O. Opoku-Agyeman[1], K. Acheampong[1], M. A. Dadzie[1],
J. Yeboah[1] and F. Owusu-Ansah[1]

[1]*Cocoa Research Institute of Ghana, P.O.Box 8, New Tafo Akim, Ghana.*

***Authors' contributions***

*This work was carried out in collaboration between all authors. All authors read and approved the final manuscript.*

*Editor(s):*
(1) Anita Biesiada, Department of Horticulture, Wroclaw University of Environmental and Life Sciences, Poland.
(2) Mintesinot Jiru, Department of Natural Sciences, Coppin State University, Baltimore, USA.
*Reviewers:*
(1) M. Monjurul Alam Mondal, Bangladesh Institute of Nuclear Agriculture, Mymensingh, Bangladesh.
(2) Anonymous, Malawi.

## ABSTRACT

Cashew cultivation in Ghana has been seriously hampered by high cost of production. This necessitated investigation into practices that will reduce establishment cost and improve field performance of cashew transplants. An experiment was conducted at Cocoa Research Institute of Ghana's (CRIG) substation at Bole (9° 01' N, 2° 29' W, altitude 309 m a s l) for optimizing the size of polybag to reduce volume of top soil required for nursing seedling, ease seedling conveyance and also improve plant establishment. Cashew seeds were sown in polybags measuring 17.5 cm x 25 cm (Large), 14.0 cm x17.8 cm (medium), 12.7 cm x 17.8 cm (small) and 10.2 cm x17.8 cm (smaller) and transplanted at 6 and 8 weeks after sowing. The experiment was laid out in a randomized complete block design with four replications. Data collected included percentage survival and growth of cashew transplants two years after transplanting and ease of seedling portage. The results showed that seedling survival was not significantly (P > 0.05) affected by the size of the polybag and age at transplanting. However bag size significantly (P < 0.001) influenced plant growth. Large polybag size produced more vigorous plants in the field. Growth of plants

---

*Corresponding author: E-mail: triciaaduy@yahoo.com;*

nursed with the medium bag sizes were also superior (P < 0.05) to the small sized bags. Seedling age did not significantly affect plant girth and height but plant leaf number was significantly (P < 0.05) affected with 8 weeks transplants producing more leaves. Seedlings in medium and small sized bags were easier to be conveyed at planting time. It is recommended that polybag sizes 14.0 cm x 17.8 cm and 12.7 cm x 17.8 cm should be used to raise cashew seedlings and transplanted at 6-weeks old to achieve higher establishment success.

*Keywords: Cashew; polybag; seedling age; transplanting survival percentage; growth.*

## 1. INTRODUCTION

Cashew (*Anacardium occidentale*) is an important non-traditional export crop in Ghana. It is a direct source of income to the farmer and a source of foreign exchange for the country. Cashew exports contributed approximately US $170 million in foreign exchange earnings to the Ghanaian economy in 2013 [1]. Cashew cultivation in Ghana began in the 1960s under the then government's savanna afforestation programme which resulted in the establishment of cashew plantations in the coastal savannah belts of the Greater Accra and the Central regions and the forest savannah transition of Brong Ahafo region [2]. In subsequent years, cashew production declined due to poor management practices and low prices. Cashew farms were subsequently abandoned despite its huge export potential. Since 1990, a renewed interest for cashew cultivation was demonstrated by farmers as a result of government's support for the industry in Ghana. This resulted in the increase in cashew cultivation and expansion of cashew farms in Ghana. Annual export of raw nuts reached 50,000 metric tonnes in 2013 [1]. In spite of this achievement, the crop is still challenged with field establishment difficulties which sometimes lead to high cost of production.

Most farms in Ghana are established either by direct seed planting or with seedlings nursed in polybags. Although direct seeding is one of the recommended field planting methods, technical advice has mainly emphasized the use of seedlings raised in polybags for establishing cashew farms because of some disadvantages associated with direct seed planting [3]. Direct seeding results in wastage of improved seeds during planting as farmers have to sow two or more seeds per hill in assurance against losses and possible mortalities [3-5]. However, in the case of seedlings nursed in polybags, the farmers have the chance to select vigorous and healthy seedlings for planting ensuring higher seedling survival and better plant growth after establishment. Seedlings may be raised in black polybags measuring 17.5 cm x 25 cm and transplanted onto the field after three months. Despite the usefulness of the polybag method, factors such as unavailability of topsoil, high cost of nursery and transportation affects polybag use [4].

The large polybags (17.5 cm x 25 cm) require approximately 3 kg of soil per bag. This size may allow about 7 to 10 seedlings to be transported by head portage per person: thus increasing time and cost of transporting seedlings for planting. Again the quantity of soil needed to fill the bags creates pressure on the limited top soil. As top soil continues to be scarce in Ghana, there is the need to find alternative polybag size to utilize less volume of soil and reduce labour and time for transporting seedlings for establishing cashew farms. Earlier work [5,6] demonstrated the feasibility of raising cashew and cocoa seedlings in smaller size bags. However the effect of the use of small size bags on establishment and plant development in the field is yet to be determined. Varying seedling age at transplanting will also determine the appropriate age to transplant cashew seedlings in small polybags to enhance survival. This study was therefore carried out to determine the effect of using small polybag sizes in nursing cashew seedlings, ease of seedling conveyance and on establishment and growth of cashew transplants in the field. It was also to determine the appropriate age to transplant the seedlings onto the field.

## 2. MATERIALS AND METHODS

The experiment was carried out at the Cocoa Research Institute of Ghana (CRIG) substation at Bole (9° 01' N, 2° 29' W, altitude 309 m above sea level) in the Northern Ghana between 2010 and 2011. Bole is in the Guinea Savannah zone of Northern Ghana with mean annual rainfall and temperature of 1087 mm and 26.1°C, respectively. The soils are mainly Ferric Luvisols with smaller areas of Eutric Regosols and Lithosols [7]. The total annual rainfalls in 2010 and 2011 were 1351.3 mm and 1132.0 mm respectively; and temperatures (min-max) were (20.9-33.2) and (20.4-32.8) during the experimental periods (source: CRIG meteorological station, Bole).

Cashew seedlings were raised in four different polybags of sizes 17.5 cm x 25 cm (large), 14.0 cm x 17.8 cm (medium), 12.7 cm x 17.8 cm (small) and 10.2 cm x 17.8 cm (smaller) at two different times in the nursery to obtain seedlings of 6 and 8 weeks old at the time of planting. The treatment combinations of polybag sizes and seedling ages were laid out in a randomized complete block design with four replications. Each treatment plot had thirty plants. The plants were spaced at 4 m x 4 m in plots measuring 24 m x 20 m.

Data collected included seedling survival (percentage), plant girth (mm), plant height (cm) and leaf number two years after field planting and the ease of transporting seedlings per person over a distance of 200 meters to the field (recorded as the average of the number of seedlings that could be carried per person over the distance). Plant survival was recorded 3 months after transplanting because after this period plant mortality may be influenced by field maintenance operations. Plant girth was measured 10 cm from the ground using a veneer caliper and plant height was recorded using a metre rule. Measurements started at planting and were repeated at 3-monthly intervals over a period of two years.

## 2.1 Data Analysis

Data were analyzed using ANOVA (GenStat 11.0 for Windows, VSN International) and treatment means compared using least significant difference (LSD) values. Data on leaf numbers was square root transformed before analysis.

## 3. RESULTS

### 3.1 Ease of Seedling Handling and Portage

The results show that handling of smaller bags was quicker than the other size bags (Table 1). Average number of bags filled per person increased with a decrease in size of bag. The number of seedlings conveyed per person by head portage to the field (200 meters) was also

higher with the smaller size bags compared to the large bags.

## 3.2 Seedling Survival

The size of bag in which the seedlings were nursed and seedling age at transplanting did not significantly (P>0.05) affect survival of cashew transplants in the field (Fig. 1). Polybag size and seedling age interaction was also not significant (P > 0.05). However seedlings transplanted at 6 weeks after sowing was observed to have higher survival than the eight weeks old seedlings after planting. Seedlings nursed with small polybag size (12.7 cm x 17.8 cm) recorded no mortalities either planted at 6 and 8 weeks after sowing.

## 3.3 Plant Girth (mm)

Polybag size significantly (P < 0.001) influenced the girth of cashew transplants two years in the field (Table 2). Plants raised in the large bag size (17.5 cm x 25 cm) had significantly (P < 0.001) bigger girths compared to those raised in the smaller bags (10.2 cm x 17.8 cm) which recorded the least girth. Seedling age at planting did not significantly (P > 0.05) influence girth of cashew transplants in the field. Similarly polybag size and seedling age interaction on plant girth was also not significant (P > 0.05).

## 3.4 Plant Height (cm)

The height of cashew transplants also showed significant differences (P < 0.001) between the polybags used two years in the field. Similar to observations on girth, plants raised in large bags (17.5 cm x 25 cm) were significantly taller, followed by medium (14.0 cm x 17.5 cm) bags which were not significantly different to those raised in the small bags (12.7 cm x 17.8 cm) (Table 3). Plants raised with the smaller bags (10.2 cm x 17.8 cm) recorded the least height. Again seedling age at transplanting did not significantly (P > 0.05) influence plant height in the field. The bag size x seedling age interaction on plant height was also not significant.

## 3.5 Plant Number of Leaves

The number of leaves produced by cashew plants after transplanting was significantly influenced by polybag sizes and seedling age at transplanting. Plants from the large bags (17.5 cm x 25 cm) produced significantly (P < 0.001) higher number of leaves when planted at 6 weeks or at 8 weeks after sowing (Table 4). Plants from the small bag size (12.7 cm x 17.8 cm) had less leaf numbers when transplanted at 6 weeks but produced more leaves when planted at 8 weeks after sowing. Averagely leaves produced by cashew plants planted at 8 weeks after sowing were significantly (P < 0.05) high compared to 6 weeks old transplants.

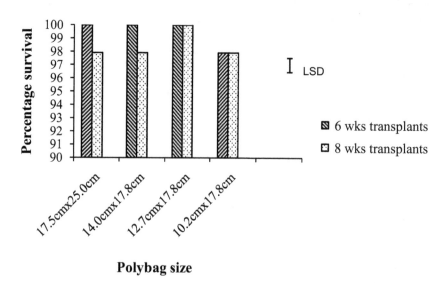

**Fig. 1. Effects of polybag size and seedling age at transplanting on plant survival**
*LSD (P > 0.05): polybag size: not significant, seedling age at transplanting: not significant, polybag size * seedling age: not significant*

### Table 1. Average number of bags filled and seedlings conveyed per person

| Polybag size | Average weight of filled bag (kg) | Average number of bags filled per person | Average number of bags carried per person |
|---|---|---|---|
| 17.5 cm x 25.0 cm | 2.6 | 400 | 10 |
| 14.0 cm x 17.8 cm | 1.0 | 800 | 25 |
| 12.7 cm x 17.8 cm | 0.8 | 1000 | 31 |
| 10.2 cm x 17.8 cm | 0.6 | 1200 | 40 |

### Table 2. Effects of polybag size and seedling age at transplanting on plant girth (mm)

| Polybag size | Plant girth (mm) | | Mean (Polybag size) |
|---|---|---|---|
| | 6 weeks transplants | 8 weeks transplants | |
| 17.5 cm x 25.0 cm | 17.7 | 17.8 | **17.7** |
| 14.0 cm x 17.5 cm | 16.7 | 16.9 | **16.8** |
| 12.7 cm x 17.8 cm | 15.9 | 16.3 | **16.1** |
| 10.2 cm x 17.8 cm | 14.9 | 15.4 | **15.1** |
| **Mean (seedling age)** | **16.3** | **16.6** | |
| LSD         : Polybag size | | 0.98** | |
|       : Seedling age | | ns | |
|       : Polybag size * seedling age | | ns | |
| CV(%)        : | | 20.4 | |

*LSD = least significant difference, CV = Coefficient of variation, ns = not significant, ** = significant at (P < 0.001)*

**Table 3. Effects of polybag size and seedling age at transplanting on plant height (cm)**

| Polybag size | Plant height (cm) | | Mean |
|---|---|---|---|
| | 6 weeks transplants | 8 weeks transplants | (polybag size) |
| 17.5 cm x 25.0 cm | 59.3 | 64.7 | **62.0** |
| 14.0 cm x 17.5 cm | 58.8 | 59.1 | **58.9** |
| 12.7 cm x 17.8 cm | 54.4 | 57.5 | **55.9** |
| 10.2 cm x 17.8 cm | 50.5 | 52.9 | **51.7** |
| **Mean (seedling age)** | **55.7** | **58.6** | |
| LSD        : Polybag size | | 4.37** | |
|             : Seedling age | | ns | |
|             : Polybag size * seedling age | | ns | |
| CV(%)      : | | 26.9 | |

*LSD = least significant difference, CV = Coefficient of variation, ns = not significant, ** = significant at (P < 0.001)*

**Table 4. Effects of polybag size and seedling age at transplanting on leaf intensity per plant**

| Polybag size | Plant number of leaves | | Mean |
|---|---|---|---|
| | 6 weeks transplant | 8 weeks transplant | (Polybag size) |
| 17.5 cm x 25 cm | 67.7 (8.1) | 68.1 (8.1) | **67.9 (8.1)** |
| 14.0 cm x 17.5 cm | 57.1 (7.5) | 60.9 (7.7) | **59.0 (7.6)** |
| 12.7 cm x 17.8 cm | 48.6 (6.9) | 64.6 (7.9) | **56.6 (7.4)** |
| 10.2 cm x 17.8 cm | 56.1 (7.3) | 57.6 (7.4) | **56.9 (7.4)** |
| **Mean (seedling age)** | **57.4 (7.4)** | **62.8 (7.8)** | |
| LSD        : Polybag size | | (0.38)** | |
|             : Seedling age | | (0.27)* | |
|             : Polybag size * seedling age | | (0.53) | |
| CV (%)     : | | 17.4 | |

*Values in parenthesis are square root transformation of the actual values. LSD = least significant difference, CV = Coefficient of variation, * = significant at (P < 0.05), ** = significant at (P < 0.001).*

## 4. DISCUSSION

Establishing farms with seedlings raised in polybags has considerable advantages. However its use by the cashew farmers' in Ghana has been low because of the invariably high cost involved in nursery care and in transporting seedlings to the field for planting [8]. The use of small polybags may be an alternative option which may be better accepted by cashew farmers because the cost of raising and transporting seedlings is low compared to large bags. It was observed in this study that, handling of the small bags was easier and less costly than the large bags. The medium to smaller polybags required less volume of soil to fill compared to the large bags. Thus about half the volume of top soil is required. More pieces of the smaller polybags could be filled in the working hours compared to the large bags. Therefore quantity of top soil and labor (man hours) required in filling the bags was also reduced. Labour and time for transporting the smaller polybags to the field was also less compared to the large polybags since more seedlings could be conveyed per person by head portage.

Seedling survival after transplanting was not significantly influenced either by bag size or seedling age at transplanting. However, it was observed that seedlings transplanted at 6 weeks after sowing tended to give slightly higher survival than 8 weeks old seedlings. Similar trends were reported in earlier studies [9,10]. This could be attributed to the observation that, at the time of transplanting many of the 8 weeks old seedlings had their roots penetrating the polybags and inevitably getting damaged during planting. This subjected those seedlings to greater transplanting shock thereby affecting establishment success. Damage to seedling tap root during transplanting has been observed as one of the main causes of transplanting failure common in older cashew seedlings [11]. It is reported [12] that shock of transplanting due to tap root damage is larger in older seedlings than smaller seedlings. Based on these observations, it would be reasonable to suggest that nursery periods of cashew seedlings raised in small polybags should not extend beyond 6 weeks. This is also an advantage since time and labour needed for nursery activities will be reduced. Seedlings raised in the small bag size (12.7 cm x

17.8 cm) were observed to have survived better which may be attributed to ease of handling of the bags. It's small size ensured proper handling which prevented the contents of the bag from falling apart and breaking the brittle roots during planting. The successful planting of the seedlings with a ball of soil around the roots may have improved survival in the field. The rapid growth of cashew transplants raised in large polybag size (17.5 cm x 25.0 cm) in the field was expected. Similar findings [13,14] were reported in mango and Indian sandalwood where larger containers produced better growth of seedlings. The relatively large volume of soil in the bag allowed the seedling roots to be exposed to more nutrients and soil moisture resulting in the initial rapid growth of seedlings which was still visible after planting in the field. It is also reported [15] that, seedlings raised in large bags have a well-developed root system contributing to better uptake of nutrient and water for vigorous plant growth. Although significant differences were observed in plant growth amongst the different polybag sizes in the field, subsequent performance cannot be predicted. The use of smaller bags is envisaged for easy adoption by many cashew farmers to enhance seedling portage and establishment.

## 5. CONCLUSION

Cashew seedlings can be raised in polybag size 14.0 cm x 17.8 cm (medium) and 12.7 cm x 17.8 cm (small) and transplanted into the field with high survival percentage. Seedlings raised in small bags are best transplanted at 6-weeks after sowing for higher establishment success. Seedling conveyance and handling was easy with the small polybags which is of benefit to the cashew farmer.

## ACKNOWLEDGEMENTS

The authors are grateful to Mr. Fredrick Daplah for providing technical assistance and all staff at Bole, substation of the Cocoa Research Institute of Ghana for their technical, contributions. This manuscript with number (CRIG/011/2015/023/001) is published with the kind permission of the Executive Director of Cocoa Research Institute of Ghana.

## COMPETING INTERESTS

Authors have declared that no competing interests exist.

## REFERENCES

1. Anonymous, Statistics from the Export Promotion Authority Ghana; 2013.

2. Addaquay J, Nyamekye-Boamah K. The Ghana cashew industry study. In Report prepared for the Ministry of Food and Agriculture (MOFA) under the Agricultural Diversification Project; 1998.

3. Adenikinju S, Esan E, Adeyemi A. Nursery techniques, propagation and management of cacao, kola, coffee, cashew and tea. Progress in Tree Crop Research in Nigeria. 2nd ed., CRIN, Ibadan, Nigeria. 1989;224.

4. Esan E. Studies on cocoa seedling (*Theobroma cacao* L.) transportation from the nursery and bare-root transplanting into the field. In Proceedings 8th International Cocoa Research Conference, Cartagena, Colombia. Cocoa Producers' Alliance; 1981.

5. Adu-Berko F, Idun I, Amoah F. Influence of the size of nursery bag on the growth and development of cashew (*Anacardium occidentale*) Seedlings. American Journal of Experimental Agriculture. 2011;1(4):440-441.

6. Oppong F, Ofori-Frimpong K, Fiakpornu R. Effect of polybag size and foliar application of urea on cocoa seedling growth. Ghana Journal of Agricultural Science. 2008;40(2):207-213.

7. FAO-UNESCO, Soil maps of the world: 1:50,000,000 Africa 6, Paris: UNESCO. 1977;299.

8. Donkor MA, Henderson CP, Jones AP. Survey to quantify adoption of CRIG recommendations. In Farming Systems Unit Research Paper 3. Cocoa Research Institute of Ghana; 1991.

9. Hammed L, Olaniyan A, Lucas E. Field Establishment of Cashew (*Anacardium occidentale* L.) Transplants as affected by nursery periods. Journal of Agricultural Science and Technology. 2012;2(11B): 1158-1164.

10. Opoku-Ameyaw K, Amoah F, Oppong F, Agene V. Determination of optimum age for transplanting cashew (*Anacardium occidentale*) seedlings in Northern Ghana. African Journal of Agricultural Research. 2007;7:296-299.

11. Deckers J, Cundall E, Shomari SH, Ngatunga A, Bassi G. Cashew. In: RH, Raemaekers (ed) Crop Production in Tropical Africa. Directorate General for

international Co-operation (DGIC) Brussel Belgium. 2001;691-700.

12. Hassan M, Rao V. Studies on the transplantations of seedlings of cashew *(Anacardium occidentale L)*. Indian Journal of Agricultural Sciences (India). 1957;27:177-184.

13. Annapurna D, Rathore TS, Joshi G. Effect of container type and size on the growth and quality of seedlings of Indian sandalwood *(Santalum album L.)*. Australian Forestry. 2004;67(2):82-87.

14. Haldankar P, Parulekar Y, Kulkarni M, Lawande K. Effect of size of polybag on survival and growth of mango grafts. Journal of Plant Studies. 2014;3(1):91.

15. Abugre S, Oti-Boateng C. Seed source variation and polybag size on early growth of Jatropha curcas. Journal of Agricultural & Biological Science. 2011;6(4).

# Permissions

The contributors of this book come from diverse backgrounds, making this book a truly international effort. This book will bring forth new frontiers with its revolutionizing research information and detailed analysis of the nascent developments around the world.

We would like to thank all the contributing authors for lending their expertise to make the book truly unique. They have played a crucial role in the development of this book. Without their invaluable contributions this book wouldn't have been possible. They have made vital efforts to compile up to date information on the varied aspects of this subject to make this book a valuable addition to the collection of many professionals and students.

This book was conceptualized with the vision of imparting up-to-date information and advanced data in this field. To ensure the same, a matchless editorial board was set up. Every individual on the board went through rigorous rounds of assessment to prove their worth. After which they invested a large part of their time researching and compiling the most relevant data for our readers.

The editorial board has been involved in producing this book since its inception. They have spent rigorous hours researching and exploring the diverse topics which have resulted in the successful publishing of this book. They have passed on their knowledge of decades through this book. To expedite this challenging task, the publisher supported the team at every step. A small team of assistant editors was also appointed to further simplify the editing procedure and attain best results for the readers.

Apart from the editorial board, the designing team has also invested a significant amount of their time in understanding the subject and creating the most relevant covers. They scrutinized every image to scout for the most suitable representation of the subject and create an appropriate cover for the book.

The publishing team has been an ardent support to the editorial, designing and production team. Their endless efforts to recruit the best for this project, has resulted in the accomplishment of this book. They are a veteran in the field of academics and their pool of knowledge is as vast as their experience in printing. Their expertise and guidance has proved useful at every step. Their uncompromising quality standards have made this book an exceptional effort. Their encouragement from time to time has been an inspiration for everyone.

The publisher and the editorial board hope that this book will prove to be a valuable piece of knowledge for researchers, students, practitioners and scholars across the globe.

# List of Contributors

**Richard Adu-Acheampong**
Entomology Division, Cocoa Research Institute of Ghana, P.O.Box 8, Tafo-Akim, Ghana

**Joseph Easmon Sarfo**
Entomology Division, Cocoa Research Institute of Ghana, P.O.Box 8, Tafo-Akim, Ghana

**Ernest Felix Appiah**
Entomology Division, Cocoa Research Institute of Ghana, P.O.Box 8, Tafo-Akim, Ghana

**Abraham Nkansah**
Entomology Division, Cocoa Research Institute of Ghana, P.O.Box 8, Tafo-Akim, Ghana

**Godfred Awudzi**
Entomology Division, Cocoa Research Institute of Ghana, P.O.Box 8, Tafo-Akim, Ghana

**Emmanuel Obeng**
Entomology Division, Cocoa Research Institute of Ghana, P.O.Box 8, Tafo-Akim, Ghana

**Phebe Tagbor**
Entomology Division, Cocoa Research Institute of Ghana, P.O.Box 8, Tafo-Akim, Ghana

**Richard Sem**
Entomology Division, Cocoa Research Institute of Ghana, P.O.Box 8, Tafo-Akim, Ghana

**L. A. F. Akinola**
Department of Animal Science and Fisheries, Faculty of Agriculture University of Port Harcourt, P.M.B 5323 Port Harcourt, Nigeria

**I. Etela**
Department of Animal Science and Fisheries, Faculty of Agriculture University of Port Harcourt, P.M.B 5323 Port Harcourt, Nigeria

**S. R. Emiero**
Department of Animal Science and Fisheries, Faculty of Agriculture University of Port Harcourt, P.M.B 5323 Port Harcourt, Nigeria

**F. I. Achuba**
Department of Biochemistry, Delta State University, PMB 1, Abraka, Nigeria

**P. N. Okoh**
Department of Biochemistry, Delta State University, PMB 1, Abraka, Nigeria

**Bárbara França Dantas**
Brazilian Corporation of Agricultural Research, Embrapa Semi-Arid, Petrolina, Pernambuco State, Brazil

**Rita de Cássia Barbosa da Silva**
Federal Rural University of Pernambuco State- UFRPE, Serra Talhada Campus, Serra Talhada, Pernambuco State, Brazil

**Renata Conduru Ribeiro**
Brazilian Corporation of Agricultural Research, Embrapa Semi-Arid, Petrolina, Pernambuco State, Brazil

**Carlos Alberto Aragão**
Department of Technology and Social Sciences- DTCS, Bahia State University, Juazeiro, Bahia State, Brazil

**Muhammad Rusdy**
Laboratory of Forage Crops and Grassland Management, Faculty of Animal Science, Hasanuddin University, Jl. Perintis Kemerdekaan, Makassar Indonesia, 90245, Indonesia

**I. J. Obare**
Kenya Plant Health Inspectorate Service, P.O.Box 49421-00100 Nairobi, Kenya

**M. G. Kinyua**
Department of Biotechnology, University of Eldoret, P.O.Box 1125-30100 Eldoret, Kenya

**O. K. Kiplagat**
Department of Biotechnology, University of Eldoret, P.O.Box 1125-30100 Eldoret, Kenya

**F. M. Mwatuni**
Kenya Plant Health Inspectorate Service, P.O.Box 49421-00100 Nairobi, Kenya

**Frank Onyemaobi Ojiako**
Department of Crop Science and Technology, Federal University of Technology, P.M.B.1526, Owerri, Imo State, Nigeria

**Sunday Ani Dialoke**
Department of Crop Science and Technology, Federal University of Technology, P.M.B.1526, Owerri, Imo State, Nigeria

**Gabriel Onyenegecha Ihejirika**
Department of Crop Science and Technology, Federal University of Technology, P.M.B.1526, Owerri, Imo State, Nigeria

**Christopher Emeka Ahuchaogu**
Department of Crop Production and Protection, Federal University Wukari, P.M.B.1020, Wukari, Taraba State, Nigeria

**Chinyere Peace Ohiri**
Department of Crop Science and Technology, Federal University of Technology, P.M.B.1526, Owerri, Imo State, Nigeria

**Omotoso Solomon Olusegun**
Department of Crop, Soil and Environmental Sciences, Ekiti State University, Ado-Ekiti, Nigeria

**K. Katono**
College of Agricultural and Environmental Sciences, Department of Crop Production, Makerere University, P.O.Box 7062, Kampala, Uganda

**T. Alicai**
National Crops Resources Research Institute (NaCRRI), Namulonge, Root Crops Programme, P.O.Box 7084, Kampala, Uganda

**Y. Baguma**
National Crops Resources Research Institute (NaCRRI), Namulonge, Root Crops Programme, P.O.Box 7084, Kampala, Uganda

**R. Edema**
College of Agricultural and Environmental Sciences, Department of Crop Production, Makerere University, P.O.Box 7062, Kampala, Uganda

**A. Bua**
National Crops Resources Research Institute (NaCRRI), Namulonge, Root Crops Programme, P.O.Box 7084, Kampala, Uganda

**C. A. Omongo**
National Crops Resources Research Institute (NaCRRI), Namulonge, Root Crops Programme, P.O.Box 7084, Kampala, Uganda

**Emmanuel Kumi**
Centre for Development Studies, Department of Social and Policy Sciences, University of Bath, Claverton Down, Bath, BA2 7A, United Kingdom
Policy and Development, School of Agriculture, University of Reading, Whiteknights, Reading, RG6 6AR, United Kingdom

**Andrew J. Daymond**
Policy and Development, School of Agriculture, University of Reading, Whiteknights, Reading, RG6 6AR, United Kingdom

**Getachew Mekonnen**
Department of Plant Sciences, College of Agriculture and Natural Resources, Mizan Tepi University, P.O.Box 260,Mizan Teferi, Ethiopia
School of Plant Sciences, College of Agriculture and Environmental Sciences, Haramaya University, P.O.Box 138, Dire Dawa, Ethiopia

**J. J. Sharma**
School of Plant Sciences, College of Agriculture and Environmental Sciences, Haramaya University, P.O.Box 138, Dire Dawa, Ethiopia

**Lisanework Negatu**
School of Plant Sciences, College of Agriculture and Environmental Sciences, Haramaya University, P.O.Box 138, Dire Dawa, Ethiopia

**Tamado Tana**
School of Plant Sciences, College of Agriculture and Environmental Sciences, Haramaya University, P.O.Box 138, Dire Dawa, Ethiopia

**S. M. H. Elseed**
Ministry of Agriculture and Animal Resources and Irrigation, Khartoum State, Sudan

**S. O. Yagoub**
Department of Agronomy, College of Agricultural Studies, Sudan University of Science and Technology Box, 73, Sudan

**I. S. Mohamed**
Department of Plant Protection, College of Agricultural Studies, Sudan University of Science and Technology, Sudan

**K. D. Tolorunse**
Department of Crop Production, School of Agriculture and Agricultural Technology, Federal University of Technology, P.M.B. 65, Minna, Niger State, Nigeria

**H. Ibrahim**
Department of Crop Production, School of Agriculture and Agricultural Technology, Federal University of Technology, P.M.B. 65, Minna, Niger State, Nigeria

**N. C. Aliyu**
Department of Crop Production, School of Agriculture and Agricultural Technology, Federal University of Technology, P.M.B. 65, Minna, Niger State, Nigeria

**J. A. Oladiran**
Department of Crop Production, School of Agriculture and Agricultural Technology, Federal University of Technology, P.M.B. 65, Minna, Niger State, Nigeria

**Patricia Adu-Yeboah**
Cocoa Research Institute of Ghana, P.O.Box 8, New Tafo
Akim, Ghana

**F. M. Amoah**
Cocoa Research Institute of Ghana, P.O.Box 8, New Tafo
Akim, Ghana

**A. O. Dwapanyin**
Cocoa Research Institute of Ghana, P.O.Box 8, New Tafo
Akim, Ghana

**K. Opoku-Ameyaw**
Cocoa Research Institute of Ghana, P.O.Box 8, New Tafo
Akim, Ghana

**M. O. Opoku-Agyeman**
Cocoa Research Institute of Ghana, P.O.Box 8, New Tafo
Akim, Ghana

**K. Acheampong**
Cocoa Research Institute of Ghana, P.O.Box 8, New Tafo
Akim, Ghana

**M. A. Dadzie**
Cocoa Research Institute of Ghana, P.O.Box 8, New Tafo
Akim, Ghana

**J. Yeboah**
Cocoa Research Institute of Ghana, P.O.Box 8, New Tafo
Akim, Ghana

**F. Owusu-Ansah**
Cocoa Research Institute of Ghana, P.O.Box 8, New Tafo
Akim, Ghana

Printed in the USA
CPSIA information can be obtained
at www.ICGtesting.com
JSHW051446221024
72173JS00006B/1597